Volume 1

The Innovation
Tools Handbook

Organizational and Operational Tools, Methods, and Techniques that Every Innovator Must Know

Volume 1

The Innovation Tools Handbook

Organizational and Operational Tools, Methods, and Techniques that Every Innovator Must Know

Edited by **H. James Harrington • Frank Voehl**

CRC Press
Taylor & Francis Group
Boca Raton London New York

CRC Press is an imprint of the
Taylor & Francis Group, an **informa** business

A PRODUCTIVITY PRESS BOOK

CRC Press
Taylor & Francis Group
6000 Broken Sound Parkway NW, Suite 300
Boca Raton, FL 33487-2742

© 2016 by Taylor & Francis Group, LLC
CRC Press is an imprint of Taylor & Francis Group, an Informa business

No claim to original U.S. Government works

Printed on acid-free paper
Version Date: 20151203

International Standard Book Number-13: 978-1-4987-6049-2 (Hardback)

Library of Congress Cataloging-in-Publication Data

Names: Harrington, H. J. (H. James), editor. | Voehl, Frank, 1946- editor.
Title: The innovation tools handbook / H. James Harrington and Frank Voehl, editors.
Description: Boca Raton, FL : CRC Press, 2016- | Includes bibliographical references and index.
Identifiers: LCCN 2015042020 | ISBN 9781498760492 (vol. 1)
Subjects: LCSH: Technological innovations--Management. | Diffusion of innovations--Management. | New products.
Classification: LCC HD45 .I53795 2016 | DDC 658.4/063--dc23
LC record available at http://lccn.loc.gov/2015042020

Visit the Taylor & Francis Web site at
http://www.taylorandfrancis.com

and the CRC Press Web site at
http://www.crcpress.com

Printed and bound in the United States of America by
Edwards Brothers Malloy on sustainably sourced paper

We are dedicating this book to Armand "Val" Feigenbaum, the father of quality-inspired innovation, and to all of the members of the International Association of Innovation Professionals (IAOIP) Tools and Methodologies Working Group/Think Tank and to the thousands of individuals who can use this book to acquaint themselves with the key tools and methodologies they need to master in order to be a truly innovative professional.

We would also like to dedicate this book to Brett Trusko, whose farsightedness and innovative thinking gave him the insight to establish the IAOIP. It is through this organization that the individuals who created this book were able to work together to write the manuscript.

Contents

Foreword

This book is part of a three-book series designed to provide its readers with the tools and methodologies that all innovators should be familiar with and able to use. These are the outputs from the Tools and Methodologies Working Group of the International Association of Innovative Professionals (IAOIP). The working group was made up of the following individuals:

- H. James Harrington, chairman
- Frank Voehl, co-chairman
- Yared Akalou
- Sifer Aseph
- Scott Benjamin
- Carl Carlson
- Richard Day
- Gul Aslan Damci
- Lisa Friedman
- Thomas Gaskin
- Dallas Goodall
- Luis Guedes
- Paul Hefner
- Dana Landry
- Elena Litovinskaia
- Chad McAllister
- Nikolaos Machairas
- Thomas Mazzone
- Pratik Mehta
- Dimis Michaelides
- Howard Moskowitz
- Michael Phillips
- Jose Carlos Arce Rioboo
- Achmad Rundi
- Max Singh
- Robert Sheesley
- Nithinart Sinthudeacha
- Henryk Stawicki
- Maria Thompson
- Hongbin Wang
- David Wheeler
- Jay van Zyl

The mission statement for the Tools and Methodology Working Group is

Using the expertise and experience of the organization's members and literature research, the working group will define the tools and methodologies that are extensively used in support of the innovation process. The working group will narrow the comprehensive list of tools and methodologies to a list of the ones that are most frequently used in the innovative process and which are the ones that innovative professionals should be confident in using effectively. For each tool and methodology, the working group will prepare a write-up that includes its definition, when it should be used, how to use it, examples of how it has been used, and a list of 5 to 15 questions that can be used to determine if an individual understands the tool or methodology.

To accomplish this mission, the working group studied the literature that was currently available to define tools and methodologies that were presently proposed or being used. They also contacted numerous universities that were teaching classes on innovation or entrepreneurship to determine what tools and methodologies they were promoting. In addition, they contacted individual consultants who are providing advice and guidance to organizations in order to identify tools and methodologies they were recommending. As a result of this research, more than 200 tools and methodologies were identified as being potential candidates for the innovative professional.

The group then sent surveys out to leading innovative lecturers, teachers, and consultants, asking them to classify each tool or methodology into one of the following categories:

- This tool or methodology is used on almost all the innovation projects = 4 points.
- This tool or methodology is used on a minimum of two out of five innovation projects = 1 point.
- This tool or methodology is seldom if ever used on innovative product projects = 0 point.
- Not familiar with the tool or methodology = −1 point.
- Never used or recommended this tool or methodology in doing innovation projects = −4 points.

We calculated the priority for each of the tools/methodologies by assigning a point value for each answer. The following are the guidelines that we followed:

- Plus four points for a tool/methodology that was always used
- Plus 1 point for a tool/methodology that is being used in at least two out of five projects
- No points for a tool/methodology that was seldom used
- Minus 1 for a tool/methodology that the expert had never heard of
- Minus 4 points for a tool/methodology that the expert never used

Our goal was to define 50 of the most effective or most frequently used tools/methodologies by the innovative practitioner. We ended up with the 76 tools/methodologies that are the most effective or the most frequently used tools/methodologies by the innovative practitioner (professional).

We then submitted the selected 76 tools/methodologies to a group of 28 practicing innovators, asking them to write a chapter on one or more of the tools/methodologies.

When we assembled the 76 chapters, we ended up with a manuscript of about 1000 pages. After a discussion with the book's editors and key people in the Tools and Methodologies Working Group, it was decided to divide the book up into the following three books:

- Creative tools/methodologies that every innovator should master
- Evolutionary or improvement tools/methodologies that every innovator should master
- Organizational/operational tools/methodologies that every innovator should master

On the basis of these three breakdowns, we went out again to innovative experts asking them to classify each tool as falling into one of the three categories. We soon realized that many of the tools were used in more than one category, so we asked the experts to classify the category that the tool is primarily used in and indicate which categories the tool/methodology could also be used in. On the basis of this study, we divided the manuscript into three books:

- *Organizational and Operational Tools, Methods, and Techniques That Every Innovator Must Know*
- *Evolutionary and Improvement Tools That Every Innovator Must Know*
- *Creative Tools, Methods, and Techniques That Every Innovator Must Know*

Each of these books contains the tools/methodologies that were rated as primarily used in that category. The results of this study can be seen in Table F.1.

Many organizations around the world are talking about how innovative they are or how they are going to improve their level of innovation. But few of them are willing to change the way they operate to encourage innovation, or are willing to take the risk necessary to be innovative. It is always safer to be followers rather than leaders. Many of them feel that it is good business to let other organizations do the research work and then react only to the technologies and products that are created by the innovative

TABLE F.1

Usage Classification for the Primary Innovative Tools and Methodologies

List of the Most Used and/or Most Effective Innovative Tools and Methodologies in Alphabetical Order

Organizational and/or operational IT&M

Evolutionary and/or improvement IT&M

Creative IT&M

	IT&M	Creative	Evolutionary	Organizational
1.	5 Why questions	S	P	S
2.	76 Standard solutions	P	S	
3.	Absence thinking	P		
4.	Affinity diagram	S	P	S
5.	Agile Innovation	S		P
6.	Attribute listing	S	P	
7.	Benchmarking		S	P
8.	Biomimicry	P	S	
9.	Brainwriting 6–3–5	S	P	S
10.	Business case development		S	P
11.	Business plan	S	S	P
12.	Cause and effect diagrams		P	S
13.	Combination methods	P	S	
14.	Comparative analysis	S	S	P
15.	Competitive analysis	S	S	P
16.	Competitive shopping		S	P
17.	Concept tree (concept map)	P	S	
18.	Consumer co-creation	P		
19.	Contingency planning		S	P
20.	CO-STAR	S	S	P
21.	Costs analysis	S	S	P
22.	Creative problem solving model	S	P	
23.	Creative thinking	P	S	
24.	Design for tools		P	
	Subtotal Number of Points	7	7	10
25.	Directed/focused/structure innovation	P	S	
26.	Elevator speech	P	S	S
27.	Ethnography	P		
28.	Financial reporting	S	S	P
29.	Flowcharting		P	S

(*Continued*)

TABLE F.1 (CONTINUED)

Usage Classification for the Primary Innovative Tools and Methodologies

List of the Most Used and/or Most Effective Innovative Tools and Methodologies in Alphabetical Order

Organizational and/or operational IT&M

Evolutionary and/or improvement IT&M

Creative IT&M

	IT&M	Creative	Evolutionary	Organizational
30.	Focus groups	S	S	P
31.	Force field analysis	S	P	
32.	Generic creativity tools	P	S	
33.	HU diagrams	P		
34.	I-TRIZ	P		
35.	Identifying and engaging stakeholders	S	S	P
36.	Imaginary brainstorming	P	S	S
37.	Innovation blueprint	P		S
38.	Innovation master plan	S	S	P
39.	Kano analysis	S	P	S
40.	Knowledge management systems	S	S	P
41.	Lead user analysis	P	S	
42.	Lotus blossom	P	S	
43.	Market research and surveys	S		P
44.	Matrix diagram	P	S	
45.	Mind mapping	P	S	S
46.	Nominal group technique	S	P	
47.	Online innovation platforms	P	S	S
48.	Open innovation	P	S	S
49.	Organizational change management	S	S	P
50.	Outcome-driven innovation	P		
	Subtotal Number of Points	15	4	7
51.	Plan–do–check–act	S	P	
52.	Potential investor present	S		P
53.	Pro-active creativity	P	S	S
54.	Project management	S	S	P
55.	Proof of concepts	P	S	
56.	Quickscore creativity test	P		

(Continued)

TABLE F.1 (CONTINUED)

Usage Classification for the Primary Innovative Tools and Methodologies

List of the Most Used and/or Most Effective Innovative Tools and Methodologies in Alphabetical Order

Organizational and/or operational IT&M

Evolutionary and/or improvement IT&M

Creative IT&M

	IT&M	Creative	Evolutionary	Organizational
57.	Reengineering/redesign		P	
58.	Reverse engineering	S	P	
59.	Robust design	S	P	
60.	S-curve model		S	P
61.	Safeguarding intellectual properties			P
62.	SCAMPER	S	P	
63.	Scenario analysis	P	S	
64.	Simulations	S	P	S
65.	Six Thinking Hats	S	P	S
66.	Social networks	S	P	
67.	Solution analysis diagrams	S	P	
68.	Statistical analysis	S	P	S
69.	Storyboarding	P	S	
70.	Systems thinking	S	S	P
71.	Synetics	P		
72.	Tree diagram	S	P	S
73.	TRIZ	P	S	S
74.	Value analysis	S	P	S
75.	Value propositions	S		P
76.	Visioning	S	S	P
	Subtotal Number of Points	7	12	7

(P) Priority Rating	Creative	Evolutionary	Organizational
Total	29	23	24

IT&M in creativity book: 29
IT&M in evolutionary book: 23
IT&M in organizational book: 24

Note: IT&M, Innovative Tools and/or Methodologies; P, primary usage; S, secondary usage; blank, not used or little used.

companies. You cannot become innovative by talking about it, changing your vision statement, or setting innovative goals. The management team must be willing to take a chance on risky projects and accept failure as a learning experiment. They need to make time for their employees to break away from the day-to-day activities to think about what could be.

The really successful innovators have changed the way they are organized and the way they operate to increase the number of ideas and the quality of ideas that are related to the organization's mission. These companies break the rules that have stifled innovation in conventional organizations. Many organizations communicate to their employees that they should take 10% to 20% of their time to focus on creating new ideas related to their work environment or the organization's product and services. Companies like Google have broken away from normal operating standards. Putting slides to get from one floor to the next is an idea that the conventional manager would think was ridiculous. Having rest areas where employees could go to meditate is an obvious waste of employees' time in the organization, as viewed by the conventional management system. Companies like Dana Corporation asked each employee to submit two ideas a month in writing. The first year they received 2,000,000 ideas and more than 80% were implemented. This included that the president was also required to submit two ideas a month related to ways to improve the way he works.

Keeping all this in mind, we felt that the first book in our three-part series on the tools and methodologies every innovator should understand and be able to use is one that addresses organizational structure and operations. Effectively and efficiently using these tools/methodologies is a key to changing the organization's culture to allow creativity to flourish.

We are presenting the following 24 tools/methodologies related to changing the organization's structure and operations.

Agile innovation	CO-STAR
Benchmarking	Costs analysis
Business case development	Financial reporting
Business plan	Focus groups
Comparative analysis	Identifying and engaging stakeholders
Competitive analysis	Innovation master plan
Competitive shopping	Knowledge management systems
Contingency planning	Market research and surveys

Organizational change management	Safeguarding intellectual properties
Potential investor presentation	Systems thinking
Project management	Value propositioning
S-curve model	Visioning

We are certainly not advocating that you need to use all 24 in order to have an innovative culture. Quite the contrary, the innovator needs to understand all 24, how they are used, and the type or results that they bring about in order to select the right combination for the specific organization they are working with.

H. James Harrington

Preface

In today's fast-moving and high-technology environment, the focus on quality has given way to a focus on innovation. Quality methodology has been shared and integrated into organizations around the world. High quality is now a given for products and services produced in Japan, the United States, Germany, Italy, China, India—yes, everywhere. Competition is more fierce and intense than ever before. Technology breakthroughs can be transferred to any part of the world in a matter of days. The people trained in schools around Shanghai are better educated than the people graduating in San Francisco or New York City. The key to being competitive is staying ahead of the competition. That means coming out faster and with more competitive products and services than the competition. The problem that every organization—be it public or private, profit or nonprofit, product or services—has is a need to have more innovative ideas effectively implemented. Today, we need to have our people generate more and better innovative ideas that can be rapidly provided to the consumer. That means that every part of the organization needs to be involved in the innovative activities. Innovative ideas cannot come from just research and development alone. We need to have more innovative processes and systems to support finance, production, sales, marketing, personnel, information technology, procurement—yes, every part of the organization. Even the person sweeping the floors can come up with an innovative idea that will drive a new product cycle. Organizations used to expect their employees, when they came to work, to stop thinking and blindly follow instructions. Today, employees need to realize that they are being paid for both their physical and mental capabilities. Our employees have to understand that if they are going to get ahead, they are now required to be more creative and more innovative at work than any place else. When they get to work, everyone needs to take off their baseball cap and put on their thinking hat in order for the organization to be successful. Today, the best worker is the best thinker, not the one who moves the most products.

Everybody is talking about the importance of innovation, how innovative they are, what innovative products they are producing, and how they need to be more innovative. Everybody is using the word *innovation* to highlight why they are different from everyone else, why customers should rely on them to provide services and product. But what does it all mean? After years of discussion, arguments, and debates, there is little agreement on what a true definition of innovation is. At one extreme, people will argue that innovation is "any new and unique idea." At the other extreme, individuals will define innovation as "a new and unique idea that is produced and delivered to an external customer who is willing to pay more for it than the cost to provide it plus a reasonable profit margin for the supplier." If you use the first definition of innovation, almost all organizations are innovative organizations. If you use the second definition of innovation, less than 5% of the organizations could be considered innovative. More than 95% of the new and unique ideas that are generated within most organizations never complete the consumer to deliverables cycle and produce a profit for the organization. My preferred definition that may or may not be in keeping with your personal beliefs is, "the process of translating an idea or invention into an intangible product, service, or process that creates value for which the consumer (the entity that uses the output from the idea) is willing to pay more than the cost to produce it."

Just so there is no confusion between innovation and creativity, creativity is defined as follows:

- *Creative*—Using the ability to make or think of new things involving the process by which new ideas, stories, products, etc., are created.
- *Create*—Make something; to bring something into existence.

The difference between creativity and innovation is that the output from the innovation has to be a value-added output, while the output from creativity does not have to be value added.

Keeping this information in mind, you can grasp the problem that the International Association of Innovation Professionals (IAOIP) was faced with when they were assigned the responsibility for (a) defining the body of knowledge for innovation, and (b) establishing a certification program for innovators. It is obvious that before you can establish a body of knowledge for innovation, you have to have an accepted definition of innovation. Moreover, before you can certify an individual as the innovator, you have to be able to define what an innovator is and does.

Again, the definition of an innovator is very debatable. The following are five different definitions of an innovator:

1. An innovator is an individual who creates a unique idea that is marketable.
2. An innovator is an individual or group that creates a unique idea and is able to guide it through the processes necessary to deliver it to the external customer.
3. An innovator is a person who has the capability to create unique ideas and has the entrepreneurship to turn these ideas into output that is marketed.
4. An innovator is a person who creates a unique idea and uses the facilities available to produce an output that is marketable.
5. An innovator is an individual who creates a unique idea that is marketable and guides it through the development process so that its value to the customer is greater than the resources required to produce it.

As you can see, some of the experts in the field define an innovator as "a person who comes up with a unique and creative idea that adds value." Other experts in the field feel that "an innovator must be capable of finding an unfulfilled need and taking it through each phase of the innovative process." This means that the innovator must be capable of

- Defining an unfulfilled need
- Creating a solution for the unfulfilled need
- Developing a value proposition
- Getting the value proposition approved by management, if it relates to an established organization
- Getting the project funded
- Establishing an organization to produce the output
- Producing the output
- Marketing the output
- Selling the output
- Evaluating the success of the project

Now this may be a lot to expect one person to be able to accomplish. But there literally are millions of these individuals successfully doing this today. You can see some excellent examples of these types of individuals on one of the most-watched new television programs called *Shark Tank*. The contestants are individuals who originated a unique idea, found ways to get it funded, and found ways to produce the output. They also set up the marketing and sales system, and sold the product. These are innovators who have come to the TV program to present their innovative output to a group of five entrepreneurs in order to get additional funding and increase sales opportunities. Basically, we believe an innovator is "a person who identifies an unfulfilled need, creates ideas that will fulfill the unmet need, and incorporate the skills of an entrepreneur."

So, what is an entrepreneur? Is an innovator also the same as an entrepreneur? Basically, an entrepreneur is someone who exercises initiative by organizing a venture to take benefit of an opportunity, and, as the decision maker, decides what, how, and how much of a good or service will be produced. An entrepreneur supplies risk capital as a risk taker, and monitors and controls the business activities. The entrepreneur is usually a sole proprietor, a partner, or the one who owns the majority of shares in an incorporated venture. (www.businessdictionary.com.)

Keeping this in mind, the real difference between an entrepreneur and an innovator is that the innovator needs to be able to recognize an unfulfilled need and create ideas that fill the unfulfilled need. The entrepreneur does not have to create the idea but can take somebody else's idea and turn it into a value-added output.

As you can see, this discussion applies very nicely to a person who is starting a new organization (e.g., a start-up company), but it is difficult to apply this to an established organization. In an established organization, the innovative process flows through many different functions, and usually the individual who recognized the unfulfilled need and created the idea to fill that unfulfilled need is not assigned to process the idea through the total innovative process. What happens in established companies is that subject matter experts are developed and assigned to individual parts of the innovative process. For example, the controller function is responsible for obtaining adequate financing; marketing defines how the unique concept will be communicated to the customer, and the sales force develops the sales campaign. Manufacturing engineering establishes the production facilities, etc. Each of these functions develops specialized skills to effectively and efficiently process the innovative concept through the innovative cycle. It is a rare exception within an established company when the individual who created the innovative idea is held responsible for having it progress through the innovative cycle and create value added to the consumer and the organization. In the cases where this occurs, that individual is called an entrepreneur. (An entrepreneur is an employee of a large corporation who is given freedom and financial support to create new products, service, systems, etc., and does not have to follow the corporation's usual routines or protocols.) In this book, we will be using the following definition of an innovator:

> An innovator is an individual who creates a unique idea that is marketable; one who then guides it through the innovative process so that its value to the customer is greater than the resources required to produce it.

Considering all the difficulties there are in getting an agreed-to definition of innovation or innovator, you might think it would be impossible to define what types of tools and methodologies should be used during the innovative cycle. Not so—we accomplished this by forming a committee to research key documentation related to books and technical articles on innovation methods and techniques. We then prepared a list of recommended tools and methodologies and distributed the list to many of the experts who were teaching or using the innovation process. Each of these experts, in turn, evaluated the long list of tools and methodologies that we had collected, and each put these tools in the following categories:

- Always used
- Frequently used
- Seldom used
- Never used
- Not known

We also asked them to suggest any additional tool/methodology that was missing from the list. As a result of this research study, we defined the tools that were most efficient, effective, and frequently used in the innovative process. These tools are presented in this book along with enough information to show you how to use them. Each tool is represented by a chapter and presented in the following format:

- Definition: tool and/or methodology.
- User: who uses the tool/methodology.

- What phases of the innovative process are the tool/methodology used in?
- How is the tool/methodology used?
- Examples of the outputs from the tool/methodology.
- Software to help in using the tool/methodology.
- References.
- Suggested additional reading.

For many of the tools, there are, or should be, complete books written on how to effectively utilize the tool/methodology. For example, there are a number of books written on storyboarding. In most cases, we have recommended additional reading for those who desire more detailed information on how to effectively implement and use the tool/methodology. We do not believe that most people involved in the innovative process will need to master all of the tools or methodologies listed in this book; however, we do believe that all of them are important enough that all the individuals involved in the innovative process should at least be familiar with each of them.

We also recommend that any actually involved in moving an innovative idea through any part of the innovative process become active members in the IAOIP. Their mission is to organize and advance innovation through the development of a catalog of innovation skills and capabilities as well as certifications to demonstrate mastery of that body of knowledge. Working groups are organized to create the base of advanced innovation certification requirements, and ensure that our certified members and organizations are global leaders in practicing and managing innovation. More information related to the IAOIP can be found at http://iaoip.org/.

H. James Harrington

Acknowledgments

We would like to acknowledge the many hours of work that each of the innovative professionals who wrote chapters for this book expended without compensation—done in order to capture and share the knowledge contained in this book. We would be remiss not to acknowledge the many long days and late evenings that Candy Rogers, Joe Mueller, and Susan Koepp-Baker put into proofreading and formatting this book. It was a major challenge to convert the creative thinking of so many individuals into a standard pattern so that the book flowed freely and logically from chapter to chapter.

This book represents the output from the Tools and Methodologies Working Group of the International Association of Innovative Professionals (IAOIP).

About the Editors

 **Dr. H. James Harrington**, chief executive officer (CEO), Harrington Management Systems. In the book *Tech Trending*, Dr. Harrington was referred to as "the quintessential tech trender." The *New York Times* referred to him as having a "... knack for synthesis and an open mind about packaging his knowledge and experience in new ways—characteristics that may matter more as prerequisites for new-economy success than technical wizardry...."

It has been said about him, "Harrington writes the books that other consultants use."

The leading Japanese author on quality, Professor Yoshio Kondo, stated: "Business Process Improvement (methodology) investigated and established by Dr. H. James Harrington and his group is some of the new strategies which brings revolutionary improvement not only in quality of products and services, but also the business processes which yield the excellent quality of the output."

The father of *total quality control*, Dr. Armand V. Feigenbaum stated: "Harrington is one of those very rare business leaders who combines outstanding inherent ability, effective management skills, broad technology background and great effectiveness in producing results. His record of accomplishment is a very long, broad and deep one that is highly and favorably recognized."

Bill Clinton, as president of the United States, appointed Dr. Harrington to serve as an Ambassador of Goodwill.

Newt Gingrich, former Speaker of the House and general chairman of American Solutions, has appointed Dr. H. James Harrington to the advisory board of his Jobs and Prosperity Task Force.

KEY RESPONSIBILITIES

H. James Harrington now serves as the CEO for the Harrington Management Systems, and he is on the board of directors for a number of small- to medium-size companies helping them develop their business strategies. He also serves as

- President of the Walter L. Hurd Foundation
- Honorary advisor for Quality for China
- Chairman of the Centre for Organizational Excellence Research (COER)
- President of the Altshuller Institute

AWARDS AND RECOGNITION

Harrington received many awards and recognition trophies throughout his 60 years of activity in promoting quality and high performance throughout the world. He has had many performance improvement awards named after him from countries around the world. Some of them are as follows:

- The Harrington/Ishikawa Medal, presented yearly by the Asian Pacific Quality Organization, was named after H. James Harrington to recognize his many contributions to the region.
- The Harrington/Neron Medal was named after H. James Harrington in 1997 for his many contributions to the quality movement in Canada.
- Harrington Best TQM Thesis Award was established in 2004 and named after H. James Harrington by the European Universities Network and e-TQM College.
- Harrington Chair in Performance Excellence was established in 2005 at the Sudan University.
- Harrington Excellence Medal was established in 2007 to recognize an individual who uses the quality tools in a superior manner.
- H. James Harrington Scholarship was established in 2011 by the ASQ Inspection Division.

PUBLICATIONS AND LECTURES

Harrington is the author of more than 40 books and hundreds of papers on performance improvement, of which more than 150 have been published in major magazines. He has given hundreds of seminars on every continent with the exception of the South Pole.

Frank Voehl, president, Strategy Associates, now serves as the chairman and president of Strategy Associates Inc. and as a senior consultant and chancellor for the Harrington Management Systems. He also serves as the chairman of the board for a number of businesses and as a Grand Master Black Belt instructor and technology advisor at the University of Central Florida in Orlando, Florida. He is recognized as one of the world leaders in applying quality measurement and Lean Six Sigma methodologies to business processes.

PREVIOUS EXPERIENCE

Frank Voehl has extensive knowledge of National Regulatory Commission, Food and Drug Administration, Good Manufacturing Practice, and National Aeronautics and Space Administration quality system requirements. He is an expert in ISO-9000, QS-9000/14000/18000, and integrated Lean Six Sigma quality system standards and processes. He has degrees from St. John's University and advanced studies at New York University (NYU), as well as an honorary doctor of divinity degree. Since 1986, he has been responsible for overseeing the implementation of quality management systems with organizations in such diverse industries as telecommunications and utilities, federal, state and local government agencies, public administration and safety, pharmaceuticals, insurance/banking, manufacturing, and institutes of higher learning. In 2002, he joined The Harrington Group as the chief operating officer (COO) and executive vice president. He has held executive management positions with Florida Power and Light and FPL Group, where he was the founding general manager and COO of QualTec Quality Services for 7 years. He has written and published/co-published more than 35 books and hundreds of technical papers on business management, quality improvement, change management, knowledge management, logistics, and teambuilding, and has received numerous awards for community leadership, service to the third world countries, and student mentoring.

CREDENTIALS

The Bahamas National Quality Award was developed in 1991 by Voehl to recognize the many contributions of companies in the Caribbean region, and he is an honorary member of its board of judges. In 1980, the City of Yonkers, New York, declared March 7 as "Frank Voehl Day," honoring him for his many contributions on behalf of thousands of youth in the city where he lived, performed volunteer work, and served as athletic director and coach of the Yonkers-Pelton Basketball Association. In 1985, he was named "Father of the Year" in Broward County, Florida. He also serves as president of the Miami Archdiocesan Council of the St. Vincent de Paul Society, whose mission is to serve the poor and needy throughout South Florida and the world.

About the Contributors

Julian Bauer is a research assistant at the University of Applied Sciences, Upper Austria. His research interests include communication and design in the innovation process as well as the topic of open innovation.

S. Ali Bokhari, PhD, is a principal at TQM College, Manchester, United Kingdom. He is a senior educational leader, business excellence specialist, author, academic, speaker, trainer, executive coach, and consultant with focus on leadership, strategy, and performance excellence with a diversified experience of more than 25 years. Dr. Bokhari earned his PhD in business and management. His professional exposure has been to manufacturing and service sector organizations. As a senior assessor for the European (EFQM) Excellence Model, he has completed several assessment projects against this model for a variety of organizations, including large companies with thousands of employees. He has successfully facilitated, led, and completed several transformation, strategic planning, quality, process, and performance improvement projects in different organizations, achieving improved organizational results. He can be contacted at dr.ali.bokhari@gmail.com.

Lisa Friedman, PhD, is a founding partner in the Enterprise Development Group (EDG), an international consulting and training firm specializing in innovation strategy and best practices. EDG is based in Palo Alto, California, in the heart of Silicon Valley, and works with innovation leaders around the world. She is the author of several innovation books and articles, is on the editorial board of the

International Journal of Innovation Science, and is a founding board member of the International Association of Innovation Professionals (IAOIP). Dr. Friedman is a frequent speaker and workshop leader on "Innovating at Startup Speed" and "Future-Driven Leadership." She also leads EDG's Silicon Valley Innovation Programs, where leaders and teams come for visits combined with integration sessions to build their learning into their own Innovation Blueprints. Participants leave with a roadmap in hand, enabling them "to take Silicon Valley home."

Herman Gyr, PhD, is a founding partner of the Enterprise Development Group and creator of the *Innovation Blueprint*, a model for engaging leaders and other stakeholders in the transformation of their enterprise. In projects ranging throughout the United States and Europe to Asia, South America, and South Africa, Dr. Gyr has specialized in working with enterprises living through periods of dramatic disruption—many of them underestimating the potential for achieving their highest aspirations during periods of change. He is co-author of *The Dynamic Enterprise: Tools for Turning Chaos into Strategy and Strategy into Action*, and is a frequently invited speaker and workshop leader around the world on strategic thinking and business transformation in the digital era.

Dana J. Landry, PhD, is a 37-year veteran of the medical device industry. He held positions in manufacturing, quality, and, for the last 20 years, in research and product development. He holds a BS, MS, MBA, and a PhD with a specialty in leadership studies. He has held certification as a new product development professional, certified quality engineer, and certified reliability engineer. He is currently an adjunct professor at New York University in the Management of Technology program. Dr. Landry was selected as Outstanding Alumni in Biomedical Engineering in 2012 by Louisiana Tech University for his career dedication to the field.

Langdon Morris leads the innovation consulting practice of InnovationLabs LLC, where he is a senior partner and co-founder. He works with a global clientele, focusing on developing and applying advanced methods in innovation and strategy to solve complex problems with very high levels of creativity. He is recognized as one of the world's leading thinkers and consultants on innovation, and his original and groundbreaking work has been adopted by corporations and universities on every continent to help them improve their innovation processes and the results they achieve.

Steven G. Parmelee is a partner at the national intellectual property law boutique Fitch, Even, Tabin & Flannery LLP. He focuses his practice on complex patent preparation and prosecution and patent portfolio management in the United States and abroad. He is an electrical engineer patent attorney with nearly 40 years of experience who has personally written nearly 2000 patent applications. He has also litigated various cases for both bench and jury trials, and has handled appellate work before the Court of Appeals for the Federal Circuit. Having served as the vice president and director of Portfolio Management at Motorola Inc., as well as many years in private practice, he brings a balanced, practical, and experienced view of intellectual property rights to the reader.

Achmad Rundi is the director of Technology and Innovation at Catalyst Global Consulting. He earned a master's degree in management of technology from NYU Polytechnic School of Engineering.

Fiona Maria Shweitzer is professor of marketing and head of the Department of Innovation Management, Design, and Industrial Design at the University of Applied Sciences, Upper Austria. Her award-winning research (three best-paper awards) focuses on open innovation and customer integration in new product development.

Simon Speller has enjoyed an extensive portfolio career, spanning 40 years in public management, academia (business school), and professional practice. Since retiring from full-time postgraduate teaching, Simon is currently a visiting lecturer with Middlesex University Business School on their MBA programs and with Birmingham University's Institute for Local Government Studies (INLOGOV).

1

Agile Innovation

Langdon Morris

CONTENTS

DEFINITION

Agile innovation is a procedure used to create a streamlined innovation process that involves everyone. If an innovation process already exists, then the procedure can be used to improve the process, resulting in a reduction of development time, resources required, costs, delays, and faults.

USER

This tool can be used by individuals, but its best use is with a group of four to eight people. Cross-functional teams usually yield the best results from this activity.

OFTEN USED IN THE FOLLOWING PHASES OF THE INNOVATIVE PROCESS

The following are the seven phases of the innovative cycle. An X after the phase name indicates that the tool/methodology is used during that specific phase.

- Creation phase X
- Value proposition phase
- Resourcing phase
- Documentation phase
- Production phase
- Sales/delivery phase
- Performance analysis phase

TOOL ACTIVITY BY PHASE

- Creation phase—During this phase, the tool is used to stimulate the creation of new ideas that will improve the product and processes.

HOW TO USE THE TOOL

The key measurements of a success of an agile innovative process would be decreases in time to market, failed product/service launches, and investment losses. With the potential for increased productivity (from improved internal processes and new perspectives), new products and services become available to the general public sooner, leading to increased revenue and reduced costs, resulting in increased profits.

Overview

Charles Darwin said it quite well:

> *In the long history of humankind (and animal kind, too), those who learned to collaborate and improvise most effectively have prevailed.*

Innovation, collaboration, and improvisation are indeed essential forces shaping all of business and all of modern life, and they have become vitally important for the individual, the organization, and indeed for all of society.

The significance of all three and their close cousin, adaptation, leads us to some essential questions:

How well are you and your company prevailing in the current environment of accelerating change?

How well positioned are you and your company to benefit from the countless new opportunities that change is bringing?

Does your organization have a rigorous innovation process?

Are you sufficiently agile to survive, and to succeed?

These questions matter so much because the scope of the challenges that every organization faces today is nothing less than enormous. Threats are everywhere, technology is accelerating, and success clearly belongs only to those rare organizations that have the capacity not only to adapt to change but also to thrive on it, and indeed to create it.

How do they do that?

The great companies disrupt industries by reshaping entire market ecosystems. The market is forced to adapt to them.

Perhaps your organization is among these exemplars?

Mastery of Innovation

The art and science of creating change is really the mastery of innovation; however, innovation takes many forms: new products and services that turn customers into evangelists; new sales channels that dominate markets and confound competitors; accelerated product development time frames that amaze and delight customers, investors, and stakeholders; and novel technologies that inspire wonder.

Agile innovation asks,

How can these outputs be achieved not just once, but consistently?

Our answer, and the essential argument of this book, can be summarized in this way:

The market is becoming brutally competitive. It is much like a war, and to survive your organization must become proficient in innovation, which of course could become one of the most powerful weapons in your arsenal. In fact, given the accelerating rate of change, *there really is no other option but innovation*.

What is required, then, to master innovation and to become an agile, adaptive, winning organization?

First, design the right business processes that enable you to out-innovate your competition, combining quality and speed.

Second, reduce the inherent risks in innovation, while making the right investment decisions in new ideas.

Third, generate ideas that are better than everyone else's by effectively engaging a larger group of people—the entire organization as well as its broader ecosystem—as effective co-ideates.

And fourth, develop and demonstrate exceptional leadership skills, because making this all happen will probably require that you provoke and lead a genuine revolution in how your organization operates.

These principles constitute agile innovation, which is a blend of great advances in technology development (agile software development) together with the leading-edge practices and principles of innovation management.

How easy to do is all this?

Well, it may not be so easy at all, and it is probably going to be a big challenge. But it is definitely achievable, and the rewards from doing so will be great.

And anyway, what is your next best alternative to innovation? Or is there even any alternative? Probably not …

Why Don't Large Firms Innovate?

Large organizations, by their very nature, have difficulty embracing a culture of continuous innovation, and they also struggle to adopt the necessary practices. However, this is neither necessary nor inevitable. The belief has to be changed, or there are going to be a lot of unpleasant consequences.

Many of the root causes that prevent large firms from innovating are related to the challenges inherent in managing a large enterprise: the need for coordination, scale, efficiency, and sustained profitability in a brutally competitive, global marketplace, and an equally brutal securities market.

Furthermore, the modern corporation is typically built on structures, rules, and processes that may also inadvertently stifle innovation, or even kill it outright. Whether intentional or not, the premature mortality of ideas and innovations occurs on a regular basis. The logic behind suppressing innovation may even make sense, in a limited, short-term context, such as when senior executives try to exercise control over unwieldy organizations so that critical objectives can be achieved.

In so doing, however, they often sacrifice important long-term benefits. Hence, one of our goals in this book is to show how short-term objectives and long-term needs can be balanced, and how operations units and innovation teams can work in close and effective partnerships rather than as bitter rivals that compete for resources.

Technology firms, like all businesses, struggle with the challenge to balance the short and the long term, to balance operational imperatives with the need to create new products and services and methods; however, one set of principles that technologists have developed that seem to be ahead of the rest are the methods widely known as *agile*.

Agile refers to a set of principles and practices that have been developed by software programming teams with two major goals: to accelerate their work and to reliably produce work of the highest quality. By reducing the burdens of bureaucratic project management, they also freed programmers to work more productively and with much greater satisfaction. This effort has been amazingly successful, and a robust methodology has emerged since *The Manifesto for Agile Software Development* (http://www .agilealliance.org) was published in the spring of 2001.

How are agile and classical innovation management related? One way to think of it is like this—agile is a speedometer, so that by knowing your speed, you can calculate when you will reach your destination. Without a speedometer or odometer, you would have to guess, and the longer and less familiar the route, the worse your prediction is going to be. Without reliable measurement, you can only guess.

Agile innovation, in complement to your agile speedometer, is the Global Positioning System (GPS) that helps you stay on course. Even without plotting your course in detail ahead of time—or developing *comprehensive documentation and specification*, you can get started right away, and if you make a wrong turn or hit a traffic jam your agile innovation GPS makes a course correction, so whether the journey is 10 miles or 10,000 miles, you always know your estimated arrival.

David versus Goliath

Many practitioners and advocates of agile tend to see themselves as Davids—small, lean, and determined, standing toe-to-toe on the battlefield and fighting an enormous bureaucracy of Goliaths, whose sole function is to suppress their creativity, innovativeness, productivity, and even the joy they find in their work.

Consequently, in much of the agile literature, there is an undertone of rebellion, and a determination to defend the capacity of programmers to self-organize because they love their work and are fully committed to performing to the utmost of their abilities.

In this respect, the agile movement reflects a profoundly important and meaningful inner drive.

Inner motivation is well known as the roots that lie behind great works of genius, while external motivation based on reward and punishment can achieve, even at its best, only compliance.

This burning desire to achieve greatness is reflected in the creation of agile, and this principle of inner drive is also a key premise of our work. By extending this same energy, the same drive, beyond a group of programmers or an R&D team to other parts of an organization, tremendous value can be created and captured.

And while our passion is to direct this marvelous innate creative force specifically toward *innovation*, the same concepts and principles can be applied in a great many aspects of your organization.

What is the story of agile innovation?

The shift we are proposing is from traditional, top–down, control-oriented management, to self-organizing, customer-driven, agile innovation. This is, of course, nothing less than a revolution—the revolution that is referred to in the title of this book.

Agile innovation entails a new way of working, and the powerful forces that are driving the economy and society will most likely require your organization to engage in this type of revolution sooner (better for you) or later (procrastinating will come with lots of adverse consequences).

Specific work processes related to project management and work team management are needed to create change and empower innovation throughout your own organization, and there are also some critical factors related to the leadership of your organization's revolution, enabling you to put your plans into action.

EXAMPLES

See the examples included in the how-to-use paragraphs just presented.

SOFTWARE

No commercial software is recommended.

SUGGESTED ADDITIONAL READING

Asaka, T. and Ozeki, K., eds. *Handbook of Quality Tools: The Japanese Approach*. Portland, OR: Productivity Press, 1998.

Harrington, H.J. and Lomax, K. *Performance Improvement Methods*. New York: McGraw-Hill, 2000.

Morris, L., Ma, M., and Wu, P.C. *Agile Innovation: The Revolutionary Approach to Accelerate Success, Inspire Engagement, and Ignite Creativity*. Hoboken, NJ: John Wiley & Sons, 2014.

2

Benchmarking

H. James Harrington

CONTENTS

Tell only the organization how bad they are, but tell the world if they are the best.

H. James Harrington

DEFINITION

- Benchmark (BMK)—the standard by which an item can be measured or judged.
- Benchmarking (BMKG)—a systematic way to identify, understand, and creatively evolve superior products, services, designs, equipment, processes, and practices to improve your organization's real performance.

USER

This tool can be used by individuals, but its best use is with a group of four to eight people. Cross-functional teams usually yield the best results from this activity.

OFTEN USED IN THE FOLLOWING PHASES OF THE INNOVATIVE PROCESS

The following are the seven phases of the innovative cycle. An X after the phase name indicates that the tool/methodology is used during that specific phase.

- Creation phase X
- Value proposition phase X
- Resourcing phase
- Documentation phase
- Production phase
- Sales/delivery phase
- Performance analysis phase X

TOOL ACTIVITY BY PHASE

- Creation phase—During this phase, benchmarking is used to provide a benchmark for the criteria that a new product/process should be able to equal or beat in order to be competitive.
- Value proposition phase—During this phase, benchmarking is often used to understand and estimate the market potential for a new product or service.
- Performance analysis phase—In this phase, benchmarking is used to compare the performance and added value of a product or service compared with the competition. The information is used to help direct future product development cycles.

How to Use the Tool

A good benchmark is one that highlights an opportunity for improvement, not one that indicates you are the best.

H. James Harrington

There are four types of benchmarking:

- Product and services
- Manufacturing processes
- Business processes
- Equipment

Darel Hall, manager of transportation and planning at AT&T's MMS division, stated that "AT&T went into benchmarking for the right reason: to improve our business and culture."

A key to be successful at this is having meaningful measurements and challenging targets. You need to understand how and why other organizations perform as they do. In addition, you need to close negative gaps between best practices and the performance of its organization's processes. All too often, the management team does not know the level of performance that is possible for the organization to reach. They do not believe that major improvements are possible. They do not know how to bring about major breakthrough improvements in their key performance indicators (KPIs). The answer to this limitation is benchmarking (Figure 2.1).

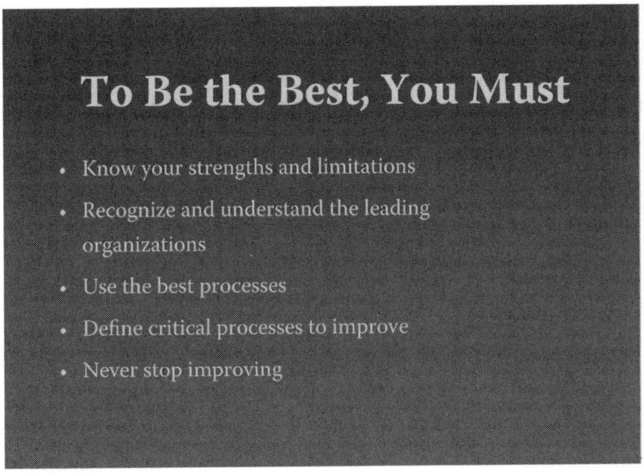

To Be the Best, You Must

- Know your strengths and limitations
- Recognize and understand the leading organizations
- Use the best processes
- Define critical processes to improve
- Never stop improving

FIGURE 2.1
To be the best.

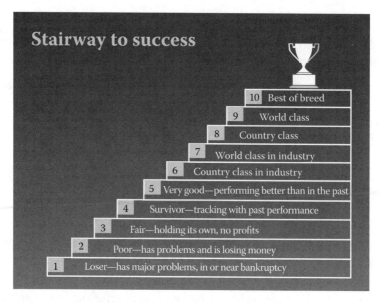

FIGURE 2.2
Stairway to success.

Figure 2.2 shows the 10 steps in benchmarking that lead to the *best of breed*. Benchmarking is a never-ending discovery and learning experience that identifies and validates the best items in order to integrate their best features into the organization's operations/products to increase their effectiveness, efficiency, and adaptability. As a result, they are able to maximize their value-added contributions to the organization. Benchmarking compares the organization and its parts to the best organization regardless of the industry or country. It helps the organization learn from the others' experiences and shows the organization how it is performing compared with the best. Benchmarking is the antidote to self-imposed mediocrity.

Benchmarking is not a new methodology (see Figure 2.3). Back in the 4th century B.C., Sun Tzu, author of *The Art of War* wrote, "If you know your enemy and know yourself, the victory will not stand in doubt." Indeed, throughout the ages, mankind has evaluated the strengths and weaknesses of others.

There are five types of benchmarking processes:

- Internal benchmarking
- External competitive benchmarking

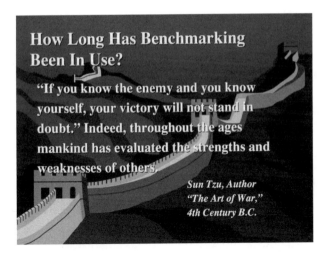

FIGURE 2.3
How long has benchmarking been in use?

- External industrial (comparable) benchmarking
- External generic (trans-industry) benchmarking
- Combined internal and external benchmarking

Figure 2.4 shows the differences in time it takes to do the benchmarking and the results that the approach gets on an average.

Comparison of the different benchmarking types

Benchmarking type	Cycle time for FSS months	Benchmarking partners	Results
Internal	3–4	Within the organization	Major improvements
External competitive	6–12	None	Better than competition
External industry	10–14	Same industry	Creative breakthrough
Combined internal and external	12–24	All industries worldwide	Best in class
External generic	12–24	All industries worldwide	Change the rules

FIGURE 2.4
Comparison of the different types of benchmarking.

If AT&T had not been into quality, I'm not sure we could have pulled off benchmarking because of the culture it needs.

Edward Trucy, Vice President, AT&T

Figure 2.5 represents a typical benchmarking process cycle. The benchmarking process today is very well defined. It is typically implemented in a five-phase process, consisting of 20 activities and 144 tasks (see Figure 2.6).

The objective of benchmarking is to define places where you can improve, not to show how good you are.

H. James Harrington

Table 2.1 provides a more detailed breakdown of the five phases and 20 activities of the benchmarking process.

It is common practice for large organizations to establish a benchmarking office. This office is a point of contact for all external organizations that want to benchmark an item within the organization (Figure 2.7).

Figure 2.8 provides insight into the time requirement in sequence of activities required to conduct a benchmarking study with an external partner. Figure 2.9 displays some comments made by Tom Carter related to benchmarking.

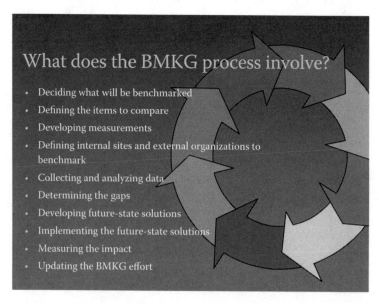

What does the BMKG process involve?

- Deciding what will be benchmarked
- Defining the items to compare
- Developing measurements
- Defining internal sites and external organizations to benchmark
- Collecting and analyzing data
- Determining the gaps
- Developing future-state solutions
- Implementing the future-state solutions
- Measuring the impact
- Updating the BMKG effort

FIGURE 2.5
BMKG circle.

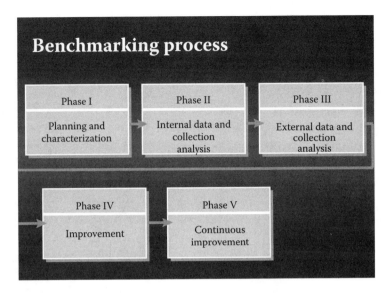

FIGURE 2.6
Five phases of the BMKG process.

TABLE 2.1

Detailed Breakdown of the Five Phases and 20 Activities of the Benchmarking Process

Benchmarking Phase	Related Activities
Phase I: Planning the benchmarking; process and characterization of the item(s)	1. Identify what to benchmark
	2. Obtain top management support
	3. Develop the measurement support
	4. Develop the data collection plan
	5. Review the plans with location experts
	6. Characterize the benchmark item
Phase II: Internal data. Collection and analysis	7. Collect and analyze internal published information
	8. Select potential internal benchmarking sites
	9. Collect internal original research information
	10. Conduct interviews and surveys
	11. Form an internal benchmarking committee
	12. Conduct internal site visits
Phase III: External data. Collection and analysis	13. Collect external published information
	14. Collect external original research information
Phase IV: Improvement of the item's performance	15. Identify corrective actions
	16. Develop an implementation plan
	17. Gain top management approval of the future-state solution
	18. Implement the future-state solution and measure its impact
Phase V: Continuous improvement	19. Maintain the benchmarking database
	20. Implement continuous performance improvement

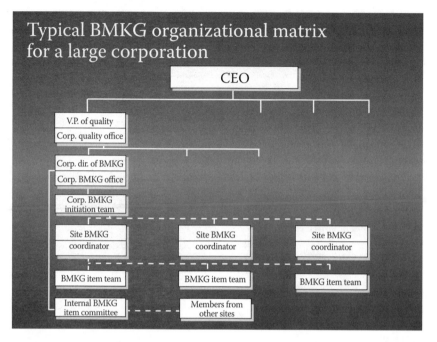

FIGURE 2.7
Typical benchmark organization matrix for a large organization.

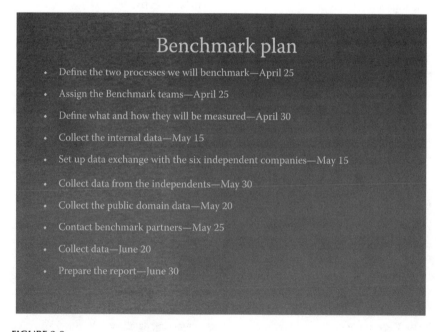

FIGURE 2.8
Time required for each part of a benchmarking process.

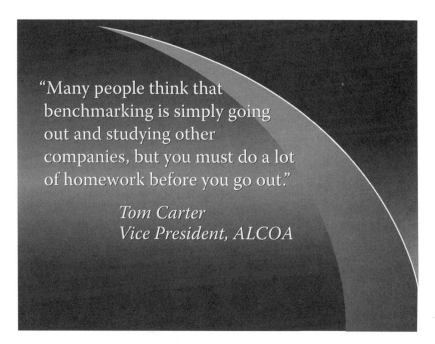

"Many people think that benchmarking is simply going out and studying other companies, but you must do a lot of homework before you go out."

Tom Carter
Vice President, ALCOA

FIGURE 2.9
Tom Carter's comments.

Many people are worried about the ethics related to the benchmarking process. The following is a typical example of the code of ethics that all organizations should subscribe to before they start benchmarking.

BENCHMARKING CODE OF CONDUCT

Introduction

Benchmarking—the process of identifying and learning from Good Practices in other organizations—is a powerful tool in the quest for continuous improvement and performance breakthroughs. Adherence to this code will contribute to efficient, effective, and ethical benchmarking.

1. Principle of preparation
    1.1 Demonstrate commitment to the efficiency and effectiveness of benchmarking by being prepared before making an initial benchmarking contact.

1.2 Make the most of your benchmarking partner's time by being fully prepared for each exchange.

1.3 Help your benchmarking partners prepare by providing them with a questionnaire and agenda before benchmarking visits.

1.4 Before any benchmarking contacts, especially the sending of questionnaires, seek legal advice.

2. Principle of contact

2.1 Respect the corporate culture of partner organizations and work within mutually agreed-on procedures.

2.2 Use benchmarking contacts designated by the partner organization if that is its preferred procedure.

2.3 Agree with the designated benchmarking contact on how communication or responsibility is to be delegated in the course of the benchmarking exercise. Check mutual understanding.

2.4 Obtain an individual's permission before providing his or her name in response to a contact request.

2.5 Avoid communicating a contact's name in an open forum without the contact's prior permission.

3. Principle of exchange

3.1 Be willing to provide the same type and level of information that you request from your benchmarking partner, provided that the principle of legality is observed.

3.2 Communicate fully and early in the relationship to clarify expectations, avoid misunderstanding, and establish mutual interest in the benchmarking exchange.

3.3 Be honest, complete, and timely with information submitted.

4. Principle of confidentiality

4.1 Treat benchmarking findings as confidential to the individuals and organizations involved. Such information must not be communicated to third parties without the prior consent of the benchmarking partner who shared the information. When seeking prior consent, make sure that you specify clearly what information is to be shared, and with whom.

4.2 An organization's participation in a study is confidential and should not be communicated externally without their prior permission.

5. Principle of use

5.1 Use information obtained through benchmarking only for purposes stated to and agreed on with the benchmarking partner.

5.2 The use of communication of a benchmarking partner's name with the data obtained or the practices observed requires the prior permission of that partner.

5.3 Contact lists or other contact information provided by benchmarking networks in any form may not be used for purposes other than benchmarking.

6. Principle of legality

6.1 Take legal advice before launching any activity.

6.2 Avoid discussions or actions that could lead to or imply an interest in restraint of trade, market or customer allocation schemes, price fixing, dealing arrangements, bid rigging, or bribery. Do not discuss costs with competitors if costs are an element of pricing. Do not exchange forecasts or other information about future commercial intentions.

6.3 Refrain from the acquisition of information by any means that could be interpreted as improper, including the breach, or inducement of a breach, of any duty to maintain confidentiality.

6.4 Do not discuss, disclose, or use any confidential information that may have been obtained through improper means, or that was disclosed by another in violation of a duty to maintain confidentiality.

6.5 Do not, as a consultant, client, or otherwise, pass on benchmarking findings to another organization without first getting the permission of your benchmarking partner and without first ensuring that the data is appropriately *blinded* and anonymous so that the participants' identity are protected.

7. Principle of completion

7.1 Follow through with each commitment made to your benchmarking partner in a timely manner.

7.2 Complete a benchmarking effort to the satisfaction of all benchmarking partners as mutually agreed on.

8. Principle of understanding and agreement

8.1 Understand how your benchmarking partner would like to be treated, and treat him or her in that way.

8.2 Agree on how your partner expects you to use the information provided, and do not use it in any way that would break that agreement.

9. Benchmarking with competitors

The following guidelines apply to benchmarking with both actual and potential competitors:

- In benchmarking with actual or potential competitors, ensure compliance with competition law. Always take legal advice before benchmarking contact with actual or potential competitors and throughout the benchmarking process. If uncomfortable, do not proceed. Alternatively, negotiate and sign a specific nondisclosure agreement that will satisfy the legal counsel representing each partner.
- Do not ask competitors for sensitive data or cause the benchmarking partner to feel that he or she must provide such data to keep the process going.
- Do not ask competitors for data outside the agreed-on scope of the study.
- Consider using an experienced and reputable third party to assemble and *blind* competitive data.
- Any information obtained from a benchmarking partner should be treated as internal, privileged communication. If *confidential* or proprietary material is to be exchanged, then a specific agreement should be executed to specify the content of the material that needs to be protected, the duration of the period of protection, the conditions for permitting access to the material, and the specific handling requirements that are necessary for that material.

BENCHMARKING PROTOCOL

Benchmarkers:

- Know and abide by the Benchmarking Code of Conduct for that part of the world.
- Have basic knowledge of benchmarking and follow a benchmarking process.
- Before initiating contact with potential benchmarking partners, determine what to benchmark, identify key performance variables to study, recognize superior performing companies, and complete a rigorous self-assessment.
- Prepare a questionnaire and fully developed interview guide, and share these in advance, if requested.

- Possess the authority to share and be willing to share information with benchmarking partners.
- Work through a specified contact and mutually agreed-on arrangements.

When the benchmarking process proceeds to a face-to-face site visit, the following behaviors are encouraged:

- Provide meeting agenda in advance.
- Be professional, honest, courteous, and prompt.
- Introduce all attendees and explain why they are present.
- Adhere to the agenda.
- Use language that is universal; do not use jargon.
- Be sure that neither party is sharing proprietary or confidential information unless prior approval has been obtained by both parties, from the proper authority.
- Share information about your own process, and if asked, consider sharing study results.
- Offer to facilitate a future reciprocal visit.
- Conclude meetings and visits on schedule.
- Thank your benchmarking partner for sharing his or her process.

All employees of the organization who are engaged in any benchmarking activities usually are required to read and agree to follow this code of conduct.

The three keys to success are

1. Have meaningful measurements and challenging targets
2. Understand how and why others perform as they do
3. Close all negative gaps

The management's problems are

- They do not know what level of performance is possible.
- They do not believe that major improvements are possible.
- They do not know how to bring about major breakthrough improvements within the organization.

Benchmarking is the answer to all three of these problems.

EXAMPLES

> A good benchmark tells you how much you need to change; benchmarking tells you how to do it.

H. James Harrington

The following is the final outcome of a benchmarking study focusing on comparing the percentage of increase in employee salaries for a specific industry.

The figures that appear in the columns represent the 5th, 25th, 75th, and 95th percentile results achieved by businesses within the sample group size. The 50th percentile represents the median point of scores within the group.

Benchmark name: Staff cost growth (%)

Description: [(staff costs − staff costs in previous year)/staff costs in previous year] × 100 [[(Q8 − Q8a)/Q8a] × 100]

This indicates the increase/decrease in staff costs of your business last year, compared with the previous to last year (see Figure 2.10).

95th percentile value = −33.15%

BM Company's actual value = 14.44%

Negative gap between Abu Dhabi Water and Electricity Authority (ADWEA) and 95th percentile = (47.59%)

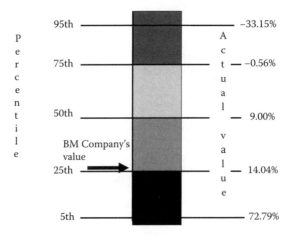

FIGURE 2.10
Comparison of percent increase in staff costs during one year.

BM Company's input data:

- A. Staff cost (2009) = 1,540,918,000 AED
- B. Staff cost (2009) = 1,346,464,000 AED

$$\left[\frac{A-B}{B}\right] \times 100 = \left[\frac{1,194,454,000}{1,346,464,000}\right] \times 100 = 14.44\%$$

Comments: BM Company is very close to the 25% percentile value. This indicates that the best organizations are reducing their cost to service the average customer. BM Company's total cost is going up to some degree because it is serving more customers (see Table 2.2).

The top 9 of the top 10 performing countries in order are

- Trinidad: 2.296
- South Korea: 4.116
- Iceland: 4.385
- Finland: 4.394
- Germany: 4.601
- Belgium: 4.662

TABLE 2.2

System Average Interruption Frequency Index

System Average Interruption Frequency Index (SAIFI)				
EDC	2006	2007	2008	2009
Citizens' Electric Company	0.14	0.25	0.26	
Pike County Light & Power Company	1.16	0.045	0.046	
UGI Utilities Inc.	0.79	0.68	0.67	
Duquesne Light Company	0.79	0.79	0.99	
PECO Energy Company	1.35	0.99	1.04	
PPL Electric Utilities Corporation	1.27	1.11	1.05	
Wellsboro Electric Company	1.50	1.63	1.07	
Pennsylvania Power Company	1.22	1.19	1.13	
Allegheny Power	1.16	1.29	1.16	
Metropolitan Edison Company	1.73	1.63	1.35	
Pennsylvania Electric Company	1.47	1.71	1.56	
BM Company				2.78

Source: Centre for Organizational Excellence Research.

- Congo DR: 5.242
- Netherlands: 5.531
- Denmark: 6.180

SOFTWARE

Some commercial software available includes but is not limited to

- MindMap: http://www.novamind.com
- Smartdraw: http://www.smartdraw.com
- QI macros: http://www.qimacros.com

SUGGESTED ADDITIONAL READING

Asaka, T. and Ozeki, K., eds. *Handbook of Quality Tools: The Japanese Approach.* Portland, OR: Productivity Press, 1998.

Harrington, H.J. *The Complete Benchmarking Implementation Guide—Total Benchmarking Management.* New York: McGraw-Hill, 1996.

Harrington, H.J. and Harrington, J.S. *High Performance Benchmarking—20 Steps to Success.* New York: McGraw-Hill, 1996.

Harrington, H.J. and Lomax, K. *Performance Improvement Methods.* New York: McGraw-Hill, 2000.

3

Business Case Development

H. James Harrington

CONTENTS

DEFINITION

A business case captures the reason for initiating a project or program. It is most often presented in a well-structured written document; however, in some cases, it also may come in the form of a short verbal agreement or presentation. The logic of the business case is that whatever resources, such as money or effort, are consumed, they should be in support of a specific business need or opportunity.

USER

This tool can be used by individuals, but its best use is with a group of four to eight people. Cross-functional teams usually yield the best results from this activity. It is best if this activity is led by an individual who is not the one who originated the business opportunity.

OFTEN USED IN THE FOLLOWING PHASES OF THE INNOVATIVE PROCESS

The following are the seven phases of the innovative cycle. An X after the phase name indicates the tool/methodology is used during that specific phase.

- Creation phase
- Value proposition phase
- Resourcing phase X
- Documentation phase X
- Production phase
- Sales/delivery phase
- Performance analysis phase X

TOOL ACTIVITY BY PHASE

- Resourcing phase—The business case is developed as part of the knowledge base that is used during the resourcing phase to determine if a project/program should have resources assigned to it. This decision is based on the projected requirements for resource consumption and the benefits the project/program are projected to accomplish compared with other ways that the resources could be utilized. This comparison is typically used to determine if the project will be included in the organization's portfolio of projects/programs, put on hold, or dropped.

- Documentation phase—The projected resources that will be consumed and the benefits derived from the project/program become key measurements that are included in the documentation.
- Performance analysis phase—During this phase, the resources consumed and the benefits gained from the completed project/program are compared to the equivalent figures included in the business case to determine if the business case objectives were met and if the project/program was successful.

HOW TO USE THE TOOL

At the January 2014 Innovation Conference put on by the International Association of Innovation Professionals held in New York City, speaker after speaker pointed out the problems related to the high percentage of initiatives that failed to produce the desired results. The data presented ranged all the way from 60% to 90% of the projects undertaken that failed to meet desired performance levels. This results in billions of dollars in unnecessary waste and lost revenue every year by companies worldwide. I personally estimate that there is more to be saved by addressing this problem than can be saved by applying Six Sigma or any of the improvement methods to other processes within the organization. We contend that the best time to stop an unsuccessful project/initiative is before it started. This can be best accomplished through the effective use of comprehensive and realistic business case analyses on the proposed project/initiative. If 90% of the new projects fail, then only 10% of them are successful. Through the effective use of a stringent business case analysis, an organization should be able to reduce by 20% to 40% the number of projects/initiatives that will not produce the desired results. If the failed project rate was reduced by only 20%, the success rate would be increased from 10% to 28%: {100% − [90% − (90% × 20%)]}. That would result in 180% increase in the organizations' innovation output. This would result in faster technology development around the world and some organizations saving billions of dollars each year.

> *A 20% reduction in project/program failure rate could increase in organizations innovative capabilities by 180%.*

A compelling business case adequately captures both the quantifiable and intangible characteristics of a proposed project. As such, business cases can range from comprehensive and highly structured, as required by formal project management methodologies, to informal and brief. Generally, the highly structured, comprehensive business cases are developed when a large amount of resources are involved in the initiative or when an initiative has a high impact upon the organization.

Information included in a formal business case could be the background of the project, the scope, the expected business benefits, the options considered (with reasons for rejecting or carrying forward each option), expected costs of the project, a gap analysis, and the expected risks. Consideration should also be given to the option of doing nothing, including the costs and risks of inactivity. From this information, the justification for the project is derived.

Business cases are created to help the management decision makers ensure that the following five general outcomes are achieved:

1. The proposed initiative will have value and relative priority compared with alternative initiatives and choices.
2. The organization has the capability to deliver the benefits.
3. Dedicated resources are working on the highest value-generating opportunities.
4. Projects with interdependencies are undertaken in the optimum sequence and are coordinated.
5. Performance of initiatives is monitored objectively—on the basis of the objectives and expected benefits laid out in the business case—and integration and lessons learned are made part of the knowledge management system.

Process to Create a Business Case

The process to create a business case consists of eight major activities and 35 tasks (see Table 3.1).

Although there are four potential inputs to the business case process (value propositions, research evaluations, proposals without value proposition, and business case preparation activities outputs that were rejected

TABLE 3.1

Eight Activities and 35 Tasks That Make Up the Business Case Process

A. Input—Value propositions
B. Input—Research evaluations
C. Input—Proposals without value propositions
D. Input—Business case preparation activities' outputs

Activity 1—Set the Proposal Context and Stimulus
- Task 1.1 Create the BCD team
- Task 1.2 Preparing the BCD team
- Task 1.3 Analysis of proposed project's input documents
- Task 1.4 Does the proposal meet the required ground rules to prepare a business case?
- Task 1.5 If the answer to task 1.4 is no, then take appropriate action

Activity 2—Define the Sponsor's Role and Test Alignment to Organizational Objectives
- Task 2.1 Define the business case sponsor's role in the BCD process
- Task 2.2 Align the project/initiative with strategic goals and objectives

Activity 3—Prepare the BCD Team's Charter and Output
- Task 3.1 Develop the BCD team's charter
- Task 3.2 Define the business case final report

Activity 4—Patent, Trademark, Copyright Considerations
- Task 4.1 Is the idea/concept an original idea/concept?
- Task 4.2 Start patent/copyright process

Activity 5—Collect Relevant Information/Data
- Task 5.1 Characterizing the current state
- Task 5.2 Characterizing proposed future state
- Task 5.3 Define the proposed future state assumptions
- Task 5.4 Define the implementation process
- Task 5.5 Define the major parameters related to the proposal
- Task 5.6 Define the quality and type of data to be collected and prioritized
- Task 5.7 Develop the data collection plan
- Task 5.8 Collecting process/product installation-related data or information

Activity 6—Projected Improvement Analysis
- Task 6.1 Characterize the current state of the parameters identified in the tasks defined in Chapter 5
- Task 6.2 Estimate the degree of change that will be brought about as a result of the project for each of the affected parameters
- Task 6.3 Compare the estimated degree of change to the projected by the individual or group that originated the project
- Task 6.4 Determine if the improvement justifies continuing the analysis

(Continued)

TABLE 3.1 (CONTINUED)

Eight Activities and 35 Tasks That Make Up the Business Case Process

Activity 7—Defining Return on Investment
- Task 7.1 Develop an estimate of resources and cycle time required to implement the proposed project
- Task 7.2 Perform sensitivity, safety, and risk analyses, and develop mitigation plans
- Task 7.3 Calculate value added to the organization
- Task 7.4 Develop proposed project/initiative recommendations

Activity 8—Presenting the Business Case to the Executive Committee
- Task 8.1 Prepare the business case final report
- Task 8.2 Set up a meeting with the executive team
- Task 8.3 Present findings and recommendations to the executive team
- Task 8.4 Was the project/initiative approved?
- Task 8.5 The project/initiative is approved
- Task 8.6 The proposal is rejected
- Task 8.7 Prepare an initial project mission statement
- Task 8.8 Closure of the BCD team

for more information or put on hold), the major input is value propositions. Often there is confusion between the difference in value proposition and business cases. On occasion, in order to reduce development cycle time, the two of them are combined together; however, this is not good practice and often leads to higher risk related to potential project failures. The following is the definition of the value proposition:

Definition: A *value proposition* is a document that defines the net benefits that will result from the implementation of a change or the use of an output as viewed by one or more of the organization's stakeholders. A value proposition can apply to an entire organization, parts thereof, or customer accounts or product or service or internal processes.

A business case captures the reasoning for initiating a project, program, or task. The business case builds on the basic information included in the value proposition by expanding it to focus on total organizational impact and the qualification of the accuracy of the data and estimates used in the analysis and recommendations. Often outcomes are expressed in worst case, best case, and most probable values. Often the final decision is made upon combinations of worst-case conditions, while the budgeting is usually based on best-case conditions. (This is sometimes referred to as minimum, maximum, and average values.) The objective of the business case evaluation is not to get the project/initiative approved or disapproved. The objective is to fairly evaluate, with a high degree of confidence, the value-added properties the proposed project/initiative will have on the organization's stakeholders. Then, on the basis of this evaluation, it recommends the management action they should take related to the proposed project/initiative.

The business case is most often presented in a well-structured written document, but in some cases may also come in the form of a short verbal argument or presentation. The logic of the business case is that whenever resources such as money or effort are consumed, they should be in support of a specific business need. A compelling business case adequately captures both the quantifiable and intangible characteristics of a proposed project. As such, business cases can range from comprehensive and highly structured, as required by formal project management methodologies, to informal and brief. Information included in a formal business case could be the background of the project, the expected business benefits, the options considered (with reasons for rejecting or carrying forward each option), the expected costs of the project, a gap analysis, and the expected risks. Consideration should also be given to the option of doing nothing, including the costs and risks of inactivity. From this information, the justification for the project is derived.

Business Case Development (BCD) Team

One of the first activities in preparing a business case is for management to assign resources to do the analysis and to prepare the documentation. This team can consist of as few as one individual and as many individuals who are required to do a thorough business analysis of the potential value that the proposed project will bring to the organization. The makeup of the team will vary from assignment to assignment and organization to organization. Basically, there are three approaches to consider:

- Approach 1—The team that generated the value proposition is often assigned to also generate the business case. This approach makes the most use of the knowledge the individuals collected as they were preparing the value proposition. This team also has a good understanding of how the management team views the proposed project, both from a negative and a positive standpoint.

- Approach 2—A second option is to have a completely new set of resources assigned that have no preset attachments/commitments to the proposed project. This has the advantage of incorporating an independent view of both the positive and negative aspects of the proposed project. The disadvantage is that the new team is starting from scratch and will repeat much of the learning process that the value proposition team went through. Even though the purpose of this independent group is to have a fresh look at the proposed project, we like to include one of the key individuals who were involved in preparing the value proposition. This individual can help the team get to the right data sources, saving a great deal of time and effort.
- Approach 3—A third approach that is sometimes used is a combination of the experienced individuals who prepared the value proposition combined with a small group of individuals who will look at the proposed project with a fresh pair of eyes.

Note: The final decision related to the makeup of the BCD team is dependent on the nature and importance of the individual project. If a project will have a major impact on an organization's bottom line, approach 2 is the best answer as it minimizes the potential risk of project failure. For important projects that will have a minor impact on the total organization's performance, approach 1, where the team that prepared the value proposition also prepares the business case, is a cost-effective way to organize the team. For projects that will add significant value to the organization but are not crucial to the success of the organization, approach 3 (a combination of approaches 1 and 2) is the recommended organizational structure.

Business Case Project Sponsor

The business case project sponsor is the key link between the business case management team and the organization's top management.

Note: Usually the sponsor and the project sponsor are the same person. In large initiatives, sometimes, additional project sponsors are assigned to individual projects that are part of the initiative that the sponsor is in charge of. (For the remainder of this book, we will be assuming that the sponsor and project sponsor is the same person with the same responsibilities.) An effective sponsor owns the project and has the ultimate responsibility for seeing that the intended benefits are realized to create the value forecasted in the business case. A good project sponsor will not interfere in the day-to-day running of the project as that is the role of the BCD team leader. But the sponsor should help the BCD team leader facilitate the necessary organizational support needed to make strategic decisions and create a successful business case outcome.

Develop the BCD Team's Charter

Never set out on a trip unless you know where you want to go. This keeps you from going down some dead end.

H. James Harrington

The primary purpose of this chapter is to ensure that there is a common understanding between all of the members of the BCD team and management related to the assignment of creating a business case evaluation and documentation of the proposed project. This will be accomplished by the BCD team preparing a project charter and defining a general outline of the subjects that will be included in their final report. For them to accomplish this, they will need to become familiar with the input documentation related to the proposed project. The charter should address the key management issues listed in Table 3.2.

TABLE 3.2

Key Management Issues

Where and how will the investment fit within the organization's broader governance and oversight structure?	Describes the governance and oversight structure for the investment.
How will the project be managed and reviewed throughout its life cycle?	Describes the project management strategy for the investment.
How will the business outcomes be realized?	Describes the outcome management strategy for the investment.
How will the business risks be mitigated and managed?	Describes the risk management strategy for the investment.
How will change be managed and implemented?	Describes the change management strategy for the investment.
How will performance be measured?	Describes the performance measurement strategy for the investment.

The following are key questions that should be addressed and discussed in preparing the BCD team charter:

Opportunity/Problem Summary
- What are the critical-to-quality elements?
- What is wrong?
- Where does the problem exist?
- How big is the opportunity for improvement?
- When and where does the problem occur?
- What is the impact of the problem?
- Is the problem statement clear to all in the organization?
- Is the problem statement too long and therefore difficult to interpret?

Business Case
- Why is the project worth doing?
- Why is it important to employees?
- Why is it important to do it now?
- What are the consequences of not doing the project now?
- How will the reduction of errors, waste, and defects affect customers, the business, and the employees?
- How does it fit with the operational initiatives and targets?

Goals and Objectives
- What results are anticipated from this project?
- Are the set goals realistic and specific?
- Are they measurable, relevant, and time bound? (see SMART goal setting method)
- Are the set goals aggressive yet achievable?
- What improvement is targeted?
- How much improvement is required?
- What issues or key performance indicators are currently tracked on a continual basis?
- Is the target identified?
- Is the specification for each measure identified?

Final Business Case Report

The BCD team needs to consider a number of factors in order for them to develop a comprehensive business case final report that will provide them with the information they need to recommend to the executive team if the proposed project/initiative should be considered for incorporation in the organization's portfolio of projects.

When developing the business case, its intended audience and the complexity of the proposed project/concept should be taken into consideration. Tailoring the business case for the decision makers will benefit its advancement. Determine the best means for engaging the target audience and adapt the message to its needs and point of view.

Considerations for properly engaging the audience include the following:

- Involvement, whenever possible, of the identified target audience throughout the development of a business case
- Engagement of the decision makers early in the process so that the business case can evolve and appropriately address any of their concerns during its development
- Planning for periodic feedback sessions and checkpoints
- Presenting the information in a manner that is easily understandable and summarized in terms that all the executive team will be able to understand

To prepare the business case final report, the BCD team needs to have completed the business case document that describes a process that comprises 11 major steps:

1. Define the problem/opportunity. Make sure you include background and circumstances.
2. Define the scope and formulate facts and assumptions.
3. Evaluate the decisions made related to the considered alternatives (including the status quo, if relevant).
4. Identify quantifiable and nonquantifiable benefits and negative impact for the proposed project.
5. Estimate the magnitude of change each benefit and negative impact will have on the as-is situation, and compare it to the proposed project estimates.
6. Develop cost estimates for the proposed project.
7. Quantify how the product situation/process will change with time if no action is taken.
8. Conduct a timeline gap analysis comparing the as-is product's situation/process to the product's situation/process after the proposed project has been implemented.

9. Define risks related to the proposed project and compare them to the risks identified in the input documentation to the business case and evaluate the adequacy of the documented mitigation plans.
10. Determine if the project is a unique project or if it is one that is already covered in approved budgets.
11. Report results and recommendations.

The business case should provide evidence that the project is a good investment for both the funding partner and the other stakeholders. Funding partners and government funding programs usually require an early-stage business case before committing to the project. You will need to complete each of the above actions in order to build a strong early-stage business case. However, the depth and extent of analysis and documentation necessary to support your case will vary depending on the proposed initiative's scope, cost, impact on the organization, and associated risks.

While the business case may be presented in various formats, there are certain elements that should be included in any written document. The table of contents in Table 3.3 is a logical sequence for the business case final report. This format can be adapted to almost any project, but be sure to present the business case in a manner that will create a favorable impression on the management, funding partner, or administrator. From the following list, choose those items that will make sense to be included in your business case final report and create a check sheet with the pertinent items. For example, not every project needs a process map, but every project needs a plan! Successful business cases utilize the majority of elements from the given outline (see Table 3.3).

Brief Explanation of the Sample Document Elements

Section i. Title Page—The title page is the first impression a reader gets of a business case. Make sure it is neat and orderly, simple, balanced, and easy to read. It contains the

- Title of project
- Project
- Project's designation (number, location, etc.)
- Name of organization
- Date of approval by organization

TABLE 3.3

Sample Table of Contents for a Business Case Final Report

Section ii. Table of Contents—The table of contents lists the major headings in the business case, and the page on which each is found. Remember to number the pages in the document. While it is the last section completed, it is placed immediately following the title page.

Section 1: Overview

This is a short description of the proposed project/initiative. This should not be more than one paragraph.

Section 2: Stakeholders

This is a simple half-page section describing who will be involved in the project, and who the business case intends to serve (e.g., stakeholders) once implemented, and their contact information. Examples include project sponsor, project manager, business analyst, technical advisor/ subject matter expert (SME), actual stakeholders (regulatory, board of directors, community, shareholders, employees, etc.), and end users, including any additional state-level, county, and municipal agencies receiving products, systems support, or services via the implementation of the business case.

Section 3: Executive Summary

This is your first and most important selling tool. It is the *sales pitch* where you create the critical first impression of the project, so it is important to summarize the most important elements of the project in a concise and compelling manner. In our experience, many members of the executive team make up their minds after reading the executive summary without reading the remainder of the report. Guidelines for writing the executive summary are as follows:

a. Describe the project concisely, avoiding excess descriptive words.
b. Document the BCD team's recommendations. The following are typical recommendations that would be made:
   - Drop the project.
   - Approve the project.

- Put the project on hold until more information is collected, and then reevaluate.
- Special funding is not required as the proposed project is already included in budgeted activities.

c. Outline the most important short-term and long-term benefits and negative impacts of the project to the organization or community.
d. Outline the costs and resource requirements.
e. Define major risks and disadvantages, if any.
f. Summarize the most important reasons for recommending or rejecting the project.
g. Finalize the executive summary after the business case is completed.

It is in your best interest to limit the executive summary to two pages or fewer. Keep it free of technical descriptions and jargon. Concentrate on explaining your reasons for undertaking the project, and what the benefits will be. It may help to do a first draft before writing your business case; this could clarify the important elements of the project plan. Write the final draft after completing the plan details. Consider the following subsections for a robust executive overview:

1. Description of the project/initiative
2. Impact: explains what the proposed project is attempting to accomplish
3. Cost–benefit analysis
4. Risks/disadvantages analysis
5. Alternatives: other options considered
6. Recommendation: the recommended course of action
7. Executive summary (total cost, breakeven point, return on investment [ROI], special skills required, major benefits, and negative impacts)

Section 4: Body of the Report

4.1 Scope

This subsection summarizes the scope of the business case, and not just the scope of the proposed project. This section answers the following key questions pertinent to scope:

a. What are the specific major elements of the project?
b. What costs are included in the case? (The costs to one section in your organization, your entire organization, or multiple organizations? Are costs to customers included?)
c. What benefits are included in the case? (What are the benefits to one section in your organization, your entire organization, or multiple organizations? Are benefits to customers included?)
d. How many years of costs and benefits are included (i.e., two-year analysis period versus four-year analysis period)?
e. What is the start and end date for the period of analysis for costs and benefits (i.e., January 1, 2013 to December 31, 2020)?
f. What is not included? If elements have been identified as being out of scope, it is important to state this early on and document accordingly to avoid *scope creep.*

4.2 BCD Team Charter

This charter defines the roles and responsibilities as assigned to the BCD team and is based on the mission statement that was provided by the executive team. Typically, this charter will assign the BCD team's responsibility of evaluating the proposed project/initiative and recommend to the executive team if it should be included in the organization's portfolio of active projects.

4.3 Project Initiative Goals and Objectives

It should be clear to the reviewer how these goals relate to the goals of the funding organization. In the best-case scenario, these should align with the objectives stated in the Strategic Plan of the organization.

- Sample business goals
  1. Increase customer satisfaction
  2. Grow market share
  3. Reduce costs
  4. Reduce cycle time
  5. Improve quality
  6. Increase inventory turns
  7. Improve productivity
  8. Improve employee morale

Goals need to be SMART: Specific, Measurable, Achievable, Realistic, and Time Sensitive (see Figure 3.1).

These define the results expected as a direct consequence of the project's completion. Such hard data verifies the value of the project, and makes it a saleable part of the funding program. They include such items as

- The number of jobs created, upgraded, or retained. Job quality may be listed.
- New or retained investment in the organization.
- New partnerships or linkages created.
- Problems or issues resolved.
- Increased profitability or lower costs.
- Increased inventory turns.
- Reduced cycle time.
- Increased volume of sales.
- New products, services, or technologies to be developed or utilized.
- Barriers to growth that will be overcome.

Some BCD projects have long- and short-term objectives to support its goals. Identify these as such if it adds to the understanding of the program. For example, the short-term objective of building a new bridge may be to replace an aging structure soon to be condemned. If, in the long term, a multilane bridge near the industrial park may help attract industry to the

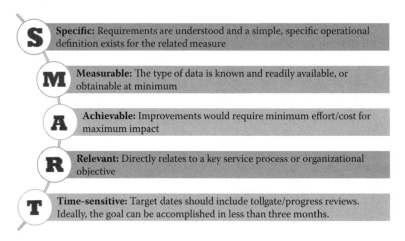

S Specific: Requirements are understood and a simple, specific operational definition exists for the related measure

M Measurable: The type of data is known and readily available, or obtainable at minimum

A Achievable: Improvements would require minimum effort/cost for maximum impact

R Relevant: Directly relates to a key service process or organizational objective

T Time-sensitive: Target dates should include tollgate/progress reviews. Ideally, the goal can be accomplished in less than three months.

FIGURE 3.1
Goals need to be SMART.

local municipality, a significant rationale will have been demonstrated. It is important that you define each goal with measurable, tangible metrics—for improvement, cost savings, customer loss/retention, or otherwise. Goals that cannot be measured—though significant intangible benefits may be a desired outcome—often are very difficult to factor when making the final decision.

Project Objectives, Measures, and Outcomes   Describe the specific project's goals in terms of objectives, measures, and outcomes. Describe performance measures as key process indicators and service targets that will be used to gauge the project's success. The goals and objectives should support the business goals and objectives defined in Section 3: Scope. These project objectives represent what precisely will be achieved by completing the project. State the objectives, measures, and outcomes clearly, and include one short/summary statement for each, in rows and columns as depicted in the example part of this chapter (see Table 3.4).

Discussion of each goal clarifies the analysis or rationale for arriving at the objective. For example: A 2012 traffic study indicated that 180,000 cars and 50,000 heavy trucks would use the new route annually, each saving five miles per trip. This would create an estimated annual savings of $1,000,000 in maintenance costs on other roads.

4.4 Problem Statement or Initiative Vision Statement

This is a concise, general statement of what the project team intends to achieve by completing the project. It explains what is to be done, why, and for whom, and helps visualize success. The vision generates momentum, builds project sponsorship, and presents clear goals to motivate and frame communications. For example: By January 2015, the town of Midland City will expand the bridge over the Industrial Hollow River to four lanes, in order to attract industry to its industrial park and to facilitate traffic flow between the industrial park, the business corridor, and the highway 295 bypass.

TABLE 3.4

Sample Project Objectives and Measures

1. Increase collections (by 200%)
2. Reduce travel time required to reach the main highway (by five minutes)
3. Reduce traffic (two-way flow) using other routes (by 25%)
4. Attract new businesses that will provide (500 new jobs)

4.5 Assumptions and Constraints

This section lists the assumptions and constraints for the options evaluated in this business case.

Note: The assumptions described in your business case should apply to all the options examined in the alternatives analysis, risk assessment, and impact analysis. Examples are provided for specific assumptions, such as

a. The project has executive support.
b. The project schedule is well detailed with recently gathered estimates from key resources.
c. Stakeholders have been identified.
d. SMEs commit to actively participate in the requirements' elicitation process.
e. Targeted resources have sufficient time to devote to project activities.
f. Project planning and analysis will address and identify potential solutions to implement.
g. Information technology (IT) resources will be available to the project on a consulting basis (if required).
h. Sufficient funding is available to complete the project.
i. The anticipated solution will successfully meet or exceed all project goals and objectives.
j. Cost of materials will not change.
k. Implementation can be completed in 1 month.
l. No major technology change will be implemented for the next three years.
m. Supplier price increases will not exceed 3% a year.
n. Interest rates will not exceed 3% per year.
o. Marketing projections are accurate to plus or minus 10%.
p. Normal productivity improvements would not exceed 5% per year.
q. Shipping costs will not increase by more than 2% a year.
r. Software development will be conducted in India at a rate that is 35% less than we would be able to do it in the United States.
s. Required vendors will be available.
t. If vendors are utilized, provisions will be made for ongoing maintenance and support.
u. The market place has vendors experienced in the type of program being implemented.

Sample constraints include the following:

a. Required adherence to standards.
b. Maximum budget.
c. Limited availability of senior/sponsor and SMEs.
d. Limited availability of IT staff and resources.
e. Limited availability of internal systems analysis and development resources.
f. Required timeline needs to be shortened.
g. Tasks that may only start once a dependency is met are given a constraint to *start no later than*.
h. With tight implementation deadlines, scope creep is a risk that must be managed on any project.

4.6 Proposed Environment

Begin by thinking of and describing the desired/proposed future state, which should also be modeled or mapped whenever possible and included in the business case under Section 17 Process Map. If software will be utilized, describe software for the project, including technical factors that may be critical to project selection, if applicable. Describe any computer, network or systems hardware, or new machinery or equipment that will be implemented for the project, including technical factors that may be critical to project selection, if applicable.

4.7 Requirements

Describe the key stakeholders' requirements. These may originate from statutory regulations or directives from executives elaborated in the project charter. If a more detailed description of technical requirements is needed to explain the business case, utilize the Requirements Matrix in Section 17. Detailed requirements describe the actions that end users perform to attain a goal, and state clearly what the new system shall deliver in terms of user experience and outcomes. Requirements should meet the following criteria:

a. Feasible/attainable
b. Consistent (without contradicting other requirements)
c. Traceable (to higher/parent level requirement)
d. Measurable and testable
e. Complete/correct

f. Unambiguous/understandable
g. Verifiable and validated
h. Necessary and relevant
i. Prioritized (high, medium, low; 3—critical, 2—important, 1—useful)
j. Independent (of implementation dependencies, other requirements)

By explaining your requirements' sources and methods, you outline how the data presented in this business case was gathered and analyzed. Requirements are gathered from various (data) sources using a variety of proven methods:

a. Current state matrix: shows the limitations of the current process or system.
b. Proposed future state matrix: lays the foundation for closing the gaps identified in the current state.
c. Pilot test results: occasionally the pilot test results yield valuable data that form the basis of key performance metrics, or a reevaluation of project goals and requirements.
d. Focus group inputs: frequently yield breakthrough requirements from the collaboration of users in the focus group. Synergistically, the users in the focus group build on and validate previously stated requirements.
e. Market studies: important if there is direct competition, or if you are looking to offer product or service innovation beyond what the completion is currently capable of delivering.

4.8 Project Milestones/Schedule

Describe the project's preliminary major milestones, deliverables, and target dates. Establish a work breakdown structure and Gantt chart using a project planning application such as Microsoft Project, and select key deliverables as milestones. Include these high-level milestones in this section, along with the projected schedule's target start and completion dates. If resource identification will be of critical importance and the resources are known (and available), it is perfectly acceptable to include a column here to identify the specific resource for a given milestone.

4.9 Benefits

Whether the goals of your business case are to reduce cycle time, lower operating costs, improve customer satisfaction levels, or help employees do their jobs better, this section needs to clearly demonstrate why we need to do it. If there are projected cost savings resulting from the project, outline it here, and demonstrate how the success of the new system will be measured. And although it should be done in measured and limited doses, we should also describe the advantages of the new system in comparison with what we have today.

4.10 Impact on Solution Selection

If there are external forces (such as state or federal mandates) that will influence the solution selection proposed by the business case, you should provide background in this section with background on these key external drivers, and set the stage for the impact analysis in the following section.

4.11 Impact Analysis

This section presents the financial and nonfinancial impacts of the options considered during the due-diligence phase of BCD, and describes the primary alternatives considered. Provide the basis for any comparable projects or benchmark models used to validate estimation data sources. Describe the estimation approach (ballpark, parametric, or detail labor and material) used.

When it comes to making a decision related to which projects/programs are included in the organization's portfolio of active projects, ROI is one of the most important considerations. As ROI is the focus of many executives, this section of the presentation should include a five-year cost–benefit analysis outlining the hard costs and savings, as well as a list of soft issues that also affect the decision. The hard costs need to include any offset to department staffs when project managers are assigned full time. Other costs include training, consulting, possible travel, software, and hardware (e.g., additional or newer-generation computers or devices may be needed). The cost offsets will increase over time as training increases. They will represent the average savings per project opportunity for your industry based on three to four projects closed per year.

Although the five-year return data will be of interest to your leadership team, the softer issues communicate the real objective—competitive advantage. Here, it is important to show the business case as an enabler of the organization's goals. Include meaningful illustrations such as *vision alignment, proven approach, fact-based management, process of continuous improvement, sustained performance,* or other selling points that fit the executive mindset. The bottom line here is not to oversell the benefits or go overboard with buzzwords, but to convince management of the project's potential to help attain the organization's predefined goals, so they can approve the project and sustain their support through inception, construction, elaboration, transition, and eventual integration into daily work.

A considerable amount of time and effort will have been invested into the business case proposal by the time it is ready for presentation. Since the purpose of the presentation is to gain buy-in, it is wise to follow a concise format and design motif, consistent with similar material the executives are accustomed to reviewing. The best approach is to use a proven presentation built on the organization's unique presentation template. Although a significant amount of time and effort will be invested into the business case proposal to get the presentation ready, an effective business case model can shave hours, if not days, from BCD time.

4.12 Funding

Use this section to list the source of the project's funding, the amount budgeted, the cost estimates (including any internal resource costs), and the net total. This should be equal or show a budget surplus if the costs are less than the budget amount allocated. This may be your last and best chance to line up additional funding if you identify cost overruns beyond what has been budgeted, allocated, or committed from your funding source.

4.13 Alternatives Analysis

This section describes the primary options evaluated, and at least one nonrecommended *alternative option* as part of due diligence. You may also want to consider including the option of not doing the project at all. A simple two-column matrix is useful for stating the reasons for selecting or rejecting each alternative. If no alternatives were available and a single-source justification is being crafted, explain this rationale here as one of the options (see Figure 3.2).

4.14 Risk Analysis

Where project risks have been identified, a risk analysis should be conducted to prioritize the risks for potential fail points. As risks are identified via brainstorming or data analysis, a risk priority number (RPN) is calculated on the basis of the potential for occurrence, severity, and

---

Preferred Option Sample

Below is a summary of the analysis of the Preferred Option. It includes a summary of financial impacts, nonfinancial impacts, and risks.

- Description of the alternative being evaluated
- Financial Impacts
  The extent of financial analysis captured in this section of the business case depends on the rigor of the financial model used. Common measures used include the following:
  - Cash flow, new cash flow, cash flow stream
  - Payback period
  - Return on Investment (ROI)
  - Discounted Cash Flow (DCF) and net present value (NPV)
  - Internal rate of return (IRR)
- Nonfinancial Impacts
  Use this section to list the positive and negative non-financial impacts of the option.
- Sensitivity and Risk
  This section identifies the likelihood of results other than the financial and non-financial impacts described above. This section should provide information on the following:
  - Sensitivity: which assumptions are most important to controlling overall results? Examples include the number of transactions or the cost of materials.
  - Risk: What is the likelihood and impact of things that could go wrong?

FIGURE 3.2
Preferred Option Sample.

current prevention controls. A list of prioritized project risks and their selected risk mitigation strategies should be included in this section. In many cases, steps may have already been taken to mitigate certain known high-level risks. By incorporating the recommended risk prevention strategies into the business case and project plan, project risks should be significantly mitigated. Closely participating in and monitoring project performance is critical for effectively guarding against the emergence of new or previously unidentified risks that may arise during the project life cycle.

4.15 Project Team Preliminary Charter

This is an abbreviated charter that will be given to the project team if the business case is approved for implementation. It should include the following:

- Scope of the problem/improvement opportunity
- Approach that the project team will use to correct the problem or to take advantage of the opportunity
- Goals and objectives for the project

- Definition of the resources that will be made available to the project
- Projected project duration
- A list of project concerns or roadblocks that have been defined

4.16 Summary

This is a short summary of the business plan document. It summarizes the recommendations that the BCD team made to the executive committee and the reasons why these recommendations were made.

Section 5: Appendices

5.1 Appendix A—Definitions

A simple glossary of terms and corresponding list of definitions is a helpful reference for nontechnical readers and those unfamiliar with certain terms utilized throughout the business case. Keep in mind that the understanding of terms, acronyms, and concepts you take for granted may be unfamiliar to your audience, and require a simple definition of common usage in the business case final report.

5.2 Appendix B—Process Map—Desired Future State (Proposed)

The old saying "a picture is worth a thousand words" applies to BCD as well. In many cases, including a flowchart, *swim lane* process map or design model provides an excellent opportunity to reduce the amount of text needed to describe the workflow of your projected future state, or the transitions resulting from user–data interactions within a system or network design.

5.3 Appendix C—Requirements Matrix

A listing of functional requirements (what the system shall do) and their priorities should be provided, as well as any supplemental requirements such as usability, performance, and security. If feasible, consider creating a traceability matrix to ensure there are no gaps in fulfilling stakeholder requests. This also allows for targeted regression testing of detailed use cases, should any top-level requirements change.

5.4 Appendix D—Decision Criteria and Methodology

This section provides the criteria for evaluating the options in the impact analysis.

- Options considered
- Preferred options
  1. Short- and long-term financial comparisons
  2. Short- and long-term nonfinancial comparisons
  3. Sensitivity and risk comparisons
  4. ROI comparisons
  5. Cultural impact comparisons
  6. Organizational change management activities comparisons
  7. Enablers and barriers comparisons
- List of recommendations

5.5 Appendix E—Change Management

To craft an effective change management program (and ensure successful implementation of your project), consider the following key elements of a change management plan (see Figure 3.3).

The modern approach to change management is to reach the majority of stakeholders and end users being affected by the change. The technology now exists to reach virtually everyone via a variety of media.

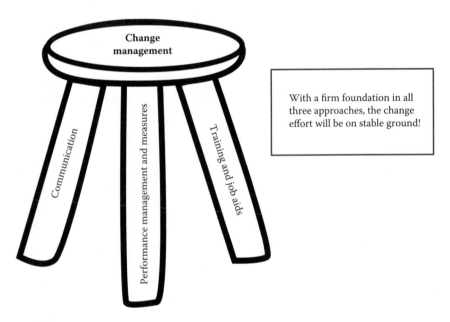

FIGURE 3.3
Key elements of a change management plan.

5.5.1 Communication   E-mail, voice, face-to-face, memos, and video messages will set the stage for change. The method of communication will depend on the demographic and level within the organization. In many cases, key stakeholders will receive change management communications from a variety of channels. For example, senior staff will receive high-level briefings, staff not likely to be affected by the change will get an e-mail, and those in the direct line of change will receive all forms of communication and go through training and receive job aids to enhance their access to real-time how-to information.

5.5.2 Training and Job Aids   Adults learn in a variety of methods and need to be reached accordingly if true change is to take hold.

A. Some are auditory learners and will review guided learning sessions with a voice-over prerecorded outlining the key components of program changes. Voice memos outlining the change and the schedule can be distributed via automated interactive voice response, as is done with many elections.
B. Some are visual learners, and will benefit from mixed-media PowerPoint presentations and videos.
C. Some are kinesthetic learners, who benefit from hands-on interaction with new systems, on-the-job training, and classroom-based training.

Some people benefit from a combination of approaches. Regardless of the media or channel, training on the new management information system will be in direct alignment with individual performance management plans.

5.5.3 Performance Management and Measures of Performance   For true change to take hold within organizations, individual performance development profiles (IDPs) need to be established and effectively implemented. The IDPs need to be aligned with departmental objectives, which in turn need to align with organizational vision, mission, and objectives. Measures of performance will be established by aligning project goals with expected behaviors. Key input, in-process, and outcome performance indicators will be collected to validate performance versus program and individual performance goals.

There are two primary obstacles to change that the BCD team is trying to overcome, by instituting comprehensive change management methods and practices within the organization:

- The first obstacle is a cultural resistance to change.

  To overcome this, people must be provided with enough information to gain an understanding of why the change is needed. A self-motivated workforce with buy-in to the change is an ideal state many organizations dream of. In lieu of this nirvana, strive to reach employees in a meaningful way by reaching out to them through a variety of media—communications via e-mail, voice, face-to-face, memos, video messages, etc.

- The second obstacle is limited resources.

  To overcome this ever-present resource limitation, the goal is to redistribute resources toward high-yield projects, by reallocating resources from resource drains—those *dead* or dying projects that were either stalled or so significantly behind schedule their chances of successful implementation were virtually nil.

Consider adopting a change management action plan to manage the transition to the new system(s).*

CHANGE MANAGEMENT ACTION PLAN

1. Plan the details of the changes, in alignment with this business case and change management action plan. Incorporate this into the project plan if possible.
2. Communicate why the change must happen, via text blast announcements, e-mail, and video messages targeted to key stakeholders.
3. Assess the impact of the change on those who will be most affected. Perform detailed risk assessment on processes, tasks, and activities targeted for elimination or automation.
4. Communicate the training requirements through learning management systems (LMSs). Test for understanding using the LMS,

---

* Adapted from *Managing Transitions—Making the Most of Change*, by William Bridges, 1991.

if possible, to also track competency and comprehension with the changes being communicated and training modules.

5. Perform targeted follow-up communications and distribute job aids to people that include an overview of the purpose of the change, a snapshot of the desired future state, step-by-step instructions for how to reach a goal within the new system, and tailor this so everyone knows—based on their role—how to do their part to facilitate successful transition to the new system(s).

6. Audit for compliance during the transition and continue collecting data on the project, and process goals and performance measures previously established. Provide real-time performance metrics so the organization and its leaders can adapt and provide real-time guidance of the change initiative(s). The auditing and monitoring can be scaled back or discontinued, once the implementation of the business case and its associated project is complete.

7. Plan for monetary and nonmonetary recognition and rewards for meeting or exceeding performance expectations and schedule a formal celebration to communicate the recognition and results.

EXAMPLES

Section A. Stakeholders
   1. Stakeholders (see Table 3.5)
Section B. Executive Overview
   2. Description, Background, and Strategic Assessment
      2.1 Executive Summary
      The Ministry of Information has been directed by the legislature of the Conch Republic to implement a national management information system (MIS). This includes security administration, tracking of enterprise data, and software application licensing. This project will fulfill the need to transition from the existing manual system to an automated system including an MIS dashboard.

      The project includes assembly of this business case to evaluate whether to

TABLE 3.5

Stakeholders

Name/Title	Role	Contact Info (E-mail)
Professor Xavier Bardo, minister of information	Project sponsor	x.bardo@moi.cr
Zia Melendez, assistant director	Project manager	z.melendez@moi.cr
Chris Voehl, business analyst	Business analyst	c.voehl@moi.cr
Henry McCoy, senior engineer	Technical advisor	h.mccoy@moi.cr
Victor Creed, systems support manager	SME	v.creed@moi.cr
Margie MacDonald, CPHP chief duty officer	End user/ stakeholder	m.macdonald@moi.cr
Ororo Monroe, Key West county training coordinator	End user/ stakeholder	o.monroe@moi.cr
Conch Republic Ministry of Information	Stakeholders	
Additional state-level, county, and municipal agencies receiving connectivity, systems support, or services via the Conch Republic Ministry of Information	End users	

   A. Hire and retain resources to completely design, develop, and administer the new system, utilizing new hardware purchased separately.

   B. Contract a vendor to design, develop, administer, and maintain the new system.

   C. Procure and implement a vendor COTS (commercial off-the-shelf) solution.

Recommendation

Option C is recommended, owing to the following factors:

- Costs related to this recommendation are similar to those of the other two options—approximately $500,000 to purchase the software and $250,000 in maintenance fees.
- Benefits—Total cost of ownership will be reduced by approximately $2,000,000 once the system is up and running. The benefit–cost ratio is approximately 2.66.
- Time to deploy is reduced with this option—less than a year.

Key Determinates

- Installing the system will save the government $2 million per year in manual process cost, real-estate leasing, overhead, storage, recycling, and disposal fees.

- Leveraging internal resources to gather requirements and perform initial system specification and process design will save costs that will go toward implementation.
- Cycle time from the general public's viewpoint will be reduced greatly depending on the particular activity. For example, the average time to get a building permit approved is six weeks. With the new system, it will be reduced to an estimated eight working days.
- Customer satisfaction/net promoter scores will improve from 62% to 80% the first year.

Cost–Benefit Analysis (see Table 3.6)

2.2 Scope

The Ministry of Information requires a system that will automate national MIS data collection and storage. The scope includes security administration, tracking of enterprise data, software application licensing, and creation of an MIS dashboard.

The primary components of the Conch Republic MIS system are computer systems hardware and software, data required for executive- and local-level decision making; design, documentation, and deployment of processes and procedures; and a human resource management system for individuals and organizations.

The implementation scope and technical implementation approach will be to implement high-priority and high-risk

TABLE 3.6

Cost–Benefit Analysis

Benefits (in Cost Savings)	
Manual process cost reduction	$750,000.00
Overhead, real estate leasing, and paper storage savings	$1,000,000.00
Recycling/disposal fee reallocation	$250,000.00
Total Benefits	$2,000,000.00
Costs	
Commercial vendor deployment estimation	$500,000.00
Implementation support, training, deployment management	$250,000.00
Internal resource time and labor costs	$50,000.00
Total Costs	$800,000.00
Net Present Value (NPV)	$1,200,000.00
Benefit–Cost Ratio	2.5

elements first, to best identify workable solutions for the majority of top-priority requirements. We will collect specific, useful, actionable information to facilitate a recommendation for the acquisition of a national MIS.

Should a COTS (commercial off-the-shelf/vendor) solution be chosen, the scope of this project will include writing a request for proposal (RFP), establishing criteria for evaluating the RFP responses, choosing the vendor, and facilitating the implementation of the solution.

The MIS system will be implemented as a web browser–based solution, either hosted externally or maintained on internal MIS servers. To ensure adequate backup and recovery, a data backup and restore process will be required and documented. Database(s) and backups of confidential data will be encrypted for security. Additional application-level security will be provided. It is anticipated that there will be thousands of end users across the Conch Republic.

After every six months of operation, the ministry will reassess the usage and storage requirements, user satisfaction levels with the system, and recommend adjustments to ensure maintainability and usability of the program.

2.3 Problem Statement

On January 1, 2013, the Conch Republic legislative body passed legislation requiring automation of manual/paper-based processes throughout the various government agencies. In addition, the current manual process has multiple opportunities for improvement, which implementation of the new system will address:

- New project approval may take 120+ days (worst-case scenario).
- Thirty-five percent of approved projects fail to be completed, due to a lack of management visibility and effective resource allocation.
- The existing manual processes are paper intensive, which results in excessive length of time, although only anecdotal data exists on the process' cycle time.
- This results in poor satisfaction levels and increased complaint calls to Congress.

– Finally, the existing manual testing process runs the risk of failing our stakeholders, resulting in potential lawsuit or disenfranchisement/defections.

3. Business Goals and Objectives

This project directly supports the strategic business goals and objectives specified in the MIS Strategic Plan 2012–2014 of Improving Performance of Key Processes by establishment of a national MIS. Combined with the stakeholder requirement to automate manual/paper-based processes, the business goals and objectives listed in Table 3.7 are the project drivers to ensure continued success of the MIS implementation.

4. Project Objectives, Measures, and Outcomes (see Table 3.8)

5. Assumptions and Constraints

v. The executive branch supports the project.

w. The project schedule is well detailed and estimates are reasonable.

x. Appropriate stakeholders are identified and commit to actively participate in requirements gathering interviews.

y. Ministry of Information, MIS Deployment Team staff, have sufficient time to devote to project activities.

z. Project planning and analysis will address and identify potential solutions to implement.

TABLE 3.7

Business Goals and Objectives

Business Goal/Objective	Description/Strategy
Goal 1: Improve stakeholder satisfaction	Key data will be gathered to calculate net promoter scores
Goal 2: Reduce costs	Cost containment analysis with all vendors to analyze preferred partner arrangements in exchange for cost reduction
Goal 3: Paperless by 2015	Reduce paper utilization, waste, and storage (file cabinets)
Goal 4: Improve visibility into key projects and processes	Establishment of a national management information system
Goal 5: Improve training and education	Improve workforce education, performance, and satisfaction
Goal 6: Improve process performance	Improve performance of key Ministry of Information processes through benchmarking, analysis, waste reduction, and metrics

TABLE 3.8

Project Objectives, Measures, and Outcomes

No.	Project Objectives	Measures/Outcomes
1.	Improved customer/stakeholder satisfaction	Net promoter scores >90%; customer satisfaction scores from 80% to >90%
2.	Cycle time for project completion and new program development	No. of hours/days/weeks reduced (total cycle time reduction)
3.	Reduce costs	Cost savings in excess of 5% of the gross domestic product
4.	Paperless processes	Reduce file cabinet/paper storage by 90%
5.	Automate key processes, including apply online for services, licensing, certification, and recertification	No. of systems automated; % of work automated; no. of users trained
6.	Implement a working COTS solution	80% of the system functionality implemented by the go-live date of January 1, 2014

aa. IT resources will be available to the project on a consulting basis.

bb. Sufficient funding is available to complete the project.

cc. The anticipated solution will include the vendor providing ongoing maintenance and support for the application.

dd. The market place has vendors experienced in MIS national level data collection systems.

ee. The procurement process will not be protested.

Constraints

i. Limited availability of Ministry of Information staff/SMEs.

j. Limited availability of IT staff.

k. Limited availability of internal systems analysis and development resources.

l. Though not anticipated, scope creep is a risk that must be managed on any project.

6. Proposed Technology Environment

6.1 Proposed software

Software Item	Description
COTS package	To be provided by selected vendor/developer

6.2 Proposed Hardware

Hardware Item	Description
Vendor or ministry hosting	The ministry will entertain either vendor-hosted or MIA-hosted solutions as long as they are consistent with the current department IT standards for security

7. Requirements

Requirement Type	Requirement Description	Priority
	Reference Appendix 2: Requirements Matrix	

8. Project Milestones/Schedule

	Milestone/Deliverable	Target Start Date	Target Completion Date
1.	Internal project initiation/scoping	Mon 6/03/13	Fri 6/28/13
2.	Elicit and document requirements	Mon 7/01/13	Fri 8/30/13
3.	Develop business case	Mon 7/29/13	Fri 8/30/13
4.	Create advertisements (RFP/ITN)	Mon 9/02/13	Mon 9/30/13
5.	Vendor contract acceptance	Tue 10/01/13	Thurs 10/31/13
6.	Vendor construction of MIS system—phase 1	Fri 11/01/13	Fri 6/13/14
7.	Testing, defect/issue reporting, resolution	Thu 1/01/14	Fri 6/13/14
8.	Change management, training, and communication	Mon 5/26/14	Fri 7/11/14
9.	System deployed	Tue 7/01/14	Fri 7/25/14
10.	Formal acceptance/celebration	Mon 7/28/14	Thu 7/31/14

SOFTWARE

Some commercial software available includes but is not limited to

- MindMap: www.novamind.com
- Smartdraw: www.smartdraw.com
- QI macros: http://www.qimacros.com

SUGGESTED ADDITIONAL READING

Harrington, H.J. and Lomax, K. *Performance Improvement Methods*. New York: McGraw-Hill, 2000.

Harrington, H.J. and Trusko, B. *Maximizing Value Proposition's to Increase Projects Success Rate*. Boca Raton, FL: CRC Press, 2014.

Voehl, C.F., Harrington, H.J., and Voehl, F. *Making the Case for Change—Using Effective Business Cases to Minimize Project and Innovation Failures*. Boca Raton, FL: CRC Press, 2014.

4

Business Plan

H. James Harrington

CONTENTS

Preparing a business plan starts the process of transforming the innovator
from a dreamer to a business executive.

H. James Harrington

DEFINITION

A business plan is a formal statement of a set of business goals, the reason they are believed to be obtainable, and the plan for reaching these goals. It also contains background information about the organization or team attempting to reach these goals.

USER

This tool is usually developed by a subset of the executive team chaired by the highest-ranking member of the organization. It is presented to the total executive team before it is formalized. It serves as the basis for all future planning and goal setting.

OFTEN USED IN THE FOLLOWING PHASES OF THE INNOVATIVE PROCESS

The following are the seven phases of the innovative cycle. An X after the phase name indicates that the tool/methodology is used during that specific phase.

- Creation phase
- Value proposition phase X
- Resourcing phase X
- Documentation phase
- Production phase
- Sales/delivery phase
- Performance analysis phase

TOOL ACTIVITY BY PHASE

- Value proposition phase—The end result of preparing the value proposition is to decide if the new concept has the potential of meeting the organization's requirements to become a committed-to activity. For those concepts that meet these requirements, the last step in the value proposition phase is preparing the initial business plan.

- Resourcing phase—There are two documents that drive the phase:
  - New concept performance specifications
  - Business plan

The *new concept performance specification* must result in output that fulfills the unmet needs of the potential consumer of the output. If this condition is not met, the concept could not be funded.

The *business plan* is analyzed to determine if the value-added content and the associated risk are in line with the expectations of the stakeholders. A negative or inadequate value-added content related to the concept should result in the project not being approved. An analysis of the key organizational managers/executives is done to determine the organization's probability of implementing the project successfully based on their education and past experience. Again, projects that an organization does not have a high probability of implementing successfully should not be funded.

- Remaining four phases of the innovation process—Throughout the remaining four phases of the innovation process, the business plan is used as a foundation for all the goal setting and planning activities. The remaining four phases are all designed to implement the concept and meet the goals and schedules as defined in the business plan. A formal business plan serves as the fundamental requirements and objectives that the innovative processes' success and/or failure is measured against.

HOW TO USE THE TOOL

Though business plans have many different presentation formats, they typically cover five major content areas. The following are the main components of a well-thought-out business plan:

- Background information for appraising your current position
- Top management and board of directors have a *strategic vision* at the core
- A marketing plan that helps you think competitively

- A summary of the operational plan, viewed as a *production system*
- A well-thought-out financial plan throughout the business plan

We recommend that all estimates be given using three cases:

- Best case
- Worst case
- Most probable case

Component One: Appraising Your Current Position

It is often told that the hardest part of initiating or appraising the current situation leading to the development of the business plan is determining an accurate assessment of your current position today. This particular section of the business plan is most appropriate for those who are developing a business plan that is being used to seek financing. Within this section, you will start by describing what stage of development your company is in and what the financing, if obtained, will be used for. Keep in mind that there are three basic reasons for seeking outside financing: (a) start-up financing, (b) expansion financing, and (c) work-out financing.

- *Start-up financing*: If your organization is in the process of seeking start-up financing, you will need to list specific milestones that have already been achieved, and then you must emphasize all positive developments without stretching the truth and being misleading. The following are some of the key questions that lenders or investors may ask:
  - Has the market research been thoroughly done? What does it show? How do you plan on using it to move forward?
  - Has a prototype product been developed or is one in the works? When will it be ready?
  - Have facilities already been leased or purchased? For how long and for how much money each month?
  - Is the management team in place? Have they invested their own money in the organization? Who are they, and what is their pedigree? Were they successful at operating other organizations? What types of contracts do they have? How about *golden parachutes*?
    - Has manufacturing been contracted for?
    - Have the initial marketing plans finalized? Who is the competition? How can they be neutralized?

 – What is the makeup of the advisory board? Have they been successful with other organizations that are at the same point in their development?

 Whether you receive financing or not and the terms of that financing will depend on the answers to the above questions and the stage of development that your organization is in. The more fully developed it is, the more favorable your financial arrangements will be.

- *Expansion financing*: If your business is already established as up and running, and you are seeking expansion financing, you need to give clear evidence that you are not merely seeking financing as a way to solve existing cash-flow problems, or to cover losses for extraordinary expenses that many times are experienced during the initial start-up.

- *Work-out financing*: Many angel investors and banks/lenders do not like to offer work-out financing, and those who are willing will want to see a business plan that clearly identifies the reasons for current or previous problems and provides a strong plan for corrective action. No matter what manner or type of financing you are seeking, lenders expect to be apprised of the source and amount of any capital that has already been secured, spent, or dissipated. They also often expect key executives to have made substantial personal equity investments in the business. They will feel even more comfortable if they recognize any other investors who may have participated in earlier stages of the financing process.

In almost every case, the first 50% of the business plan is geared toward helping develop and support a solid business strategy. Management and the board must begin by looking at the market, the industry, the customer base, and its competition. They need to analyze and study customer needs and the benefits of current products and services. Next, they need to evaluate the strengths and weaknesses of each competing organization in order to look for opportunities in the marketplace during the next year and beyond. All of these activities are largely aimed at helping senior management create an ongoing strategy for the organization and its subsidiaries, if they have any.

The second 50% of the business plan is devoted largely to executing the selected business strategy. The products and services, marketing, and operations should all closely tie in with the strategy. So while it may be easy to select a smart-sounding strategy for your plan, we recommend you give a lot of thought to the *strategic vision* that will set the course for your business. Although some form of strategic planning is critical to achieving results, we maintain that the *old* strategic planning maps and templates are no longer viable because we are largely sailing through uncharted waters in a turbulent world, making it virtually impossible to predict long-term changes in the future for a system. Answers and direction emerge, and a manager must be aware of them. In doing so, two major issues must be analyzed, evaluated, and dealt with:

1. The role of strategic vision in an unpredictable world. (For more detailed information on strategic visions, see Chapter 24.)
2. The need to contrast complexity-oriented perspectives with traditional management perspectives.

One definition of a strategic vision comes from Burt Nanus, a well-known expert on the subject. Nanus in his writings defines a strategic vision as a realistic, credible, attractive future for [an] organization.* Let us break this down as follows:

- *Realistic*: A strategic vision must always be based on reality to be meaningful for an organization, as opposed to an ordinary vision. For example, if you are a master planner charged with developing a strategic vision for a computer software company that has carved out a small niche in the marketplace developing instructional software and has a 3% share of the computer software market, a vision to overtake Microsoft and dominate the software market is not realistic, but a strategic vision might call for a 5% niche during the next 24 months.
- *Credible*: A strategic vision must be believable to be relevant. More important, it must be credible to the employees and all members of the organization. If the members of the organization do not find the vision strategically credible, it will not be able to create and harvest

---

* Burt Nanus is a well-known expert on leadership and the author of many books on the subject, including *Visionary Leadership*. Now professor emeritus of management at the University of Southern California, he was also research director of the Leadership Institute. His most often quoted phrase on the subject of strategic visioning is strategic vision (Nanus and Dobbs, 1999), "*is where tomorrow begins, for it expresses what you and others who share the vision will be working hard to create. Since most people don't take the time to think systematically about the future, those who do, and who base their strategies and actions on their visions, have inordinate power to shape the future.*"

energy, or serve a useful and meaningful purpose. Among the inherent purposes of a strategic vision are to inspire those in the organization to achieve a level of excellence, to redouble their energy, and to provide purpose and strategic direction for the employees, and to do so it must be deployable. A vision that is not credible or deployable is not strategic in its nature and will accomplish neither of these ends.

- *Attractive*: If a vision is going to strategically inspire and motivate those in the organization, it must have a strong magnetic attraction and draw people to it in order to create energy and synergistically leaven the organizational energy. In other words, people must want to be part of this future that is envisioned for the organization.

- *Future based*: A strategic vision is never in the present, and is always in the future, yet is somehow tied to the past. In this respect, the image of the leader and his planners gazing off into the distance to formulate a vision may not be too far from the truth, but can be somewhat misguided if not channeled by the core values of the organization. The paradox lies in the strategic vision, which is not where you are now but where you want to be strategically in the future— providing you do not repeat the failures of the past; otherwise, you are organizationally doomed to repeat it.

Nanus in his writings goes on to say that the right strategic vision for an organization—one that is a realistic, credible, and an attractive future for that organization—accomplishes a number of things for the organization:

- It attracts commitment and energizes people. This is one of the primary reasons for having a vision for an organization: its motivational effect. When people can see that the organization is committed to a vision, and that entails more than just having a vision statement, it generates enthusiasm about the course the organization intends to follow and increases the commitment of people to work toward achieving that vision.

- It creates meaning in workers' lives. A vision allows people to feel like they are part of a greater whole, and hence provides meaning for their work. The right vision will mean something to everyone in the organization if they can see how what they do contributes to that vision. Consider the difference between the hotel service worker who can only say, "I make beds and clean bathrooms," to the one who can also say, "I'm part of a team committed to becoming the worldwide leader in providing quality service to our hotel guests." The work is the same, but the context and meaning of the work is different.

- It establishes a standard of excellence. A vision serves a very important function in establishing a standard of excellence. In fact, a good vision is all about excellence. Tom Peters, the author of *In Search of Excellence*, talks about going into an organization where a number of problems existed. When he attempted to get the organization's leadership to address the problems, he got the defensive response, "But we're no worse than anyone else!" Peters cites this sarcastically as a great vision for an organization: "Acme Widgets: We're No Worse than Anyone Else!" A vision so characterized by lack of a striving for excellence would not motivate or excite anyone about that organization. The standard of excellence also can serve as a continuing goal and stimulate quality improvement programs, as well as providing a measure of the worth of the organization.

- It bridges the present and the future. The right vision takes the organization out of the present, and focuses it on the future. It is easy to get caught up in the crisis of the day, and to lose sight of where you were heading. A good vision can orient you on the future, and provide positive direction. The vision alone is not enough to move you from the present to the future, however. A vision is the desired future state for the organization; the strategic plan is how to get from where you are now to where you want to be in the future.

Another definition of vision comes from Oren Harari: "Vision should describe a set of ideals and priorities, a picture of the future, a sense of what makes the company special and unique, a core set of principles that the company stands for, and a broad set of compelling criteria that will help define organizational success."

A strategic vision expresses core values that inform action, and it nurtures the capabilities needed to pursue this action. It is so very much more than speculating about future trends and events, or forecasting the impact of future technologies. It is grounded in the past, while looking to the future. And because such a strategic vision is durable and long lasting, it should not require revision every few years like most ordinary visions do.

Component Two: Top Management and Board of Directors Have a Strategic Vision at the Core of the Business Plan

The foundation of every business plan must be a core set of strategic vision statements that drives the organization's present and future operations. The business plan is directed to make a step forward in the evolution from where the organization is operating or performing today toward the future vision statements developed by top management and approved by the board of directors.

Component Three: Think Competitively Throughout Your Plan

In today's highly competitive and overly crowded marketplace, you are probably going to have serious competition no matter how creative your business concept is. That is why you need to think competitively throughout your business plan. You need to realistically identify where you will do things in a similar manner as your competitors, where you will do things differently, where you have real strengths, and where you have real weaknesses.*

As part of the business planning process, you need to learn how to develop and write an effective marketing plan that answers the following questions:

1. How large is the potential market? How do you know? How is it segmented?
2. Does your customer analysis show that the market is growing, flattening, or shrinking? What data supports your conclusions?
3. How many people or businesses are currently using one of your competitor's products that are the same or similar to the one you are offering or plan to offer? How can you differentiate yours? How many customers or prospects potentially have any possible use for the product?

---

* While data is being collected, the analysis process can begin. In our research, we noticed how patterns formed from one interview to another. After the interviews were completed and transcribed, we began a formal coding system by organizing these patterns into central tendencies and ranges: (a) central tendencies describe how the data chunk together into the research participants' common themes or categories; (b) ranges allow for the differences within those categories to be discussed.

Market Segmentation

In almost every case that we have studied during the past 30 years, every market has some major and distinctive segments. Even if the marketplace is not currently segmented, the probability exists that it could or will be is real and of urgent concern. This is particularly true if the marketplace for your product or service is multiregional or international. If this is the case, segmentation is vital, especially for a small organization, if your desire is to be competitive in that your master planners will need to discuss segmentation within your business category. How you intend to cope with any positive or negative effects segmentation may have on your particular business is vital, as most marketplaces are segmented by price, features, and quality issues. Generally, however, price and quality alone do not provide the clearest or most definitive market segmentation. Much stronger segmentation can usually be found through an evaluation of product features or service uses, and its importance to various consumers.

Consumer Analysis

In your business plan, you will need to evaluate the typical consumers within the market segments you are targeting in order to determine if the market is growing, flattening, or shrinking. Although there are innumerable variables to consider when analyzing consumer behavior, you need to try to develop surveys to focus on those behavioral possibilities that best determine how viable your product will be in your target markets.*

Customer analysis starts by looking at which features will most appeal to consumers.

- How are choices made between competing products, and which marketing promotions or media avenues seem to offer the best vehicles for reaching the consumer base?
- How much disposable income do target consumers have to spend on this product?
- How do your target consumers reach purchasing decisions?
- Are consumers presold on a particular brand before they visit a store, or do they buy on impulse?

---

* What is a *survey*? The American Statistical Association discusses current practice and the innovations now in process for the main types of surveys—mail, telephone interview, and in-person interview. Each form has distinct advantages and disadvantages. Each is also greatly being changed by the use of computers, the Internet, and telecommunications.

- What characteristics influence the purchase of one product or service over a competing one?
- How many people or businesses are currently using one of your competitor's products that are the same or similar to the one you are offering or plan to offer?

You will need to include an overview of those organizations and their products, programs, and services that you will be in direct competition with. Next, try to identify the market leader and define what makes it successful and emphasize those characteristics of the firm or offerings that are different than yours. The initial reaction is to dismiss this section of the business plan simply because you may not have any real current competition. However, if there is no product or service similar to yours on the market, then you need to identify those firms that provide products or services that perform essentially the same or similar functions by making an attempt to identify any organizations that are likely to enter the market, or those competitors that are in the process of developing products or services that will be competitive with those your organization is or will be offering. It is very important to be clear not only about the distinguishing features of your program, product, or service but also to delineate any strong consumer benefits. What makes your product or service significantly better than competitive offerings?

Next, you will need to do an in-depth analysis of the competitive advantages and weaknesses of your organization and should include information that will help allay any concerns that may arise as to their ability to significantly hinder your success. This particular section of the business plan is important, especially if your company is a start-up, because you will be competing with established companies that have inherent advantages such as brand recognition, financial strength, and established distribution channels.*

---

* A broad definition of competitive analysis is the action of gathering, analyzing, and distributing information about products, customers, competitors, and any aspect of the environment needed to support executives and managers in making strategic decisions for an organization. The key points of this definition includes the following: (a) Competitive intelligence (CI) is an ethical and legal business practice. This is important as CI professionals emphasize that the discipline is not the same as industrial espionage, which is both unethical and usually illegal. (b) The focus is on the external business environment. (c) There is a process involved in gathering information, converting it into intelligence, and then utilizing this in business decision making. CI professionals emphasize that if the intelligence gathered is not usable (or actionable), then it is not intelligence. (*Source: Wikipedia*. For further information, see http://en.wikipedia.org/wiki/Competitive_intelligence.)

Additionally, strategic positioning can be thought of as a marketing strategy for your product or service in that it defines how you are going to feature your product to your targeted marketplace. The first step here is to decide who your target market will be, consisting of those potential customers toward whom you will direct most of your marketing efforts. In many cases, this group will not be the most unique or even the largest market for your product, but based on competitive factors and product benefits, it is the one that you feel your marketing arm can most effectively reach. Start-ups are generally successful if they focus on a highly specific yet very narrow target marketplace, as big general markets are usually dominated by large, well-established organizations.

Once you have determined who your target market is, you need to decide if you want consumers to perceive your product as the premium quality leader or as a low-cost substitute. Also, if you have a one-product organization or more likely a service company, your marketing strategy may coincide with your overall business strategy. This does not necessarily have to be the case; however, it is important that your product strategy be in sync with your overall business strategy.

Advertising and promotion activities are used to provide an overview of your general promotional plan, and you need to provide a summary of what methods and media you intend to use and why.* Although an advertising slogan or unique selling proposition is not strictly necessary, you should outline the proposed mix of your advertising media, use of publicity, and/or other promotional programs as follows:

- Explain how your choice of marketing strategies and tools will allow you to reach your target market.
- Explain how they will enable you to best convey your product features and benefits.

---

* Measurement, Statistics, and Methodological Studies (MSMS)—College of Education, Arizona State University (USA). On their website, you will find Dr. B's Wide World of Web Data with links to hundreds of online data sets all over the world that students and professors can use for in-class work. The Data Gallery provides a large number of graphic depictions of data. The Equation Gallery describes and presents various formulas for the statistics in one and in two dimensions. In the Graphic Studio, you can experiment with data entry forms that produce plots in HTML. In the Computing Studio, you can type data on a form that generates a page that walks you through the activities of computing your statistics. And just for the fun of it, you can get a random statistical quote from a collection of quotes from the history, philosophy, and practice of data analysis and statistics. It might stimulate your reflection on common statistical practice. Editor: John Behrens.

- Be sure that your advertising, publicity, and promotional programs sound realistic, based on your proposed marketing budget.
- Effective advertising generally relies on message repetition in order to motivate consumers to make a purchase.
- If you are on a limited budget, it is better to reach fewer, more likely prospects than too many people occasionally.

Concerning your sales strategy, it needs to be in harmony with your business strategy, marketing strategy, and your company's SWOTs (strengths and weaknesses, opportunities, and threats). If your start-up company is planning on selling products to other businesses in a highly competitive marketplace, your marketplace entrance unveiling will be easier if you rely on wholesalers or commissioned sales representatives who already have created and formed an established presence and reputation in the marketplace. On the other hand, if your business is to sell high-technology products with a range of customized options, your sales force needs to be extremely knowledgeable and personable.

Your executives need to remember that banks generally lend money to businesses on a short-term basis, and many venture capitalists may want to cash out in a few years.

Component Four: Operating Plan as a Production System

Think of your business plan as a production system with the inputs (what goes in at the start) as raw material and unfinished assemblies. In our case, the raw materials include

- Talent and initiative from your employees
- Capital
- Market position
- The company's creditworthiness
- The organization's earning capacity
- Assessment of changes in the marketplace

The structure of the business plan is unique and, in many cases, like nothing you have seen before. It is a nuts-and-bolts document, how-to-get-it-done-without-fail tool kit designed to grab your organization and take it where it needs to go. You would not find much of the theoretical conceptualizing and strategizing so many other planning books carry on about. That works for larger companies that can wait five years or more for results. Instead, what happens to your organization during the next 12 months is of major concern to your investors, for if you can hit your short-range targets year after year, the longer term will take care of itself.

In the business plan, operations is a catch-all term used to describe any important aspects of the business not described elsewhere, and should be covered in an *annual operations plan*. If the start-up is a manufacturing organization, you will need to discuss critical elements of the manufacturing process. For business plans covering retail businesses, you need to discuss store operations, while wholesalers discuss warehouse operations. In addition to discussing areas that are critical to operations, briefly summarize how major business functions will be carried out, and how certain functions may run more effectively than those of your competitors.

If there are any key vital positions that have not been filled, you will need to describe position responsibilities and the type of employment experience necessary to the position. If there is a board of directors, introduce each member and provide a concise summary profile of each person's background. Also, if they will have an active role in running your business organization, you should elaborate on that role. Finally, if consultants have been engaged for key responsibilities, include a brief description of their backgrounds and functions, and how they will be engaged and their exit strategy. Finally, fill as many of your key positions as possible before you seek funding. Many financiers reject plans if the management team is incomplete.

Component Five: Well-Thought-Out Financial Plan

In this section, you need to show projected, or *pro forma*, income statements, balance sheets, and cash flow. Existing businesses should also show historical financial statements. While how far into the future you need to project and the number of possible scenarios you can anticipate depends on the complexity of the business, three to five years for financial projections and three scenarios are average. Scenarios should be based on the most likely course your business will take, a weak scenario with sales coming in well under expectation, and a good scenario with projected sales well over expectation.

Pro-forma income statements should show sales, cost of operation, and profits on both a monthly and annual basis for each plan year. For all but the largest businesses, annual pro-forma balance sheets are all that are necessary. Cash flow pro formas should be presented in both monthly and annual form. And, if your business is already established, past annual balance sheets and income statements should also be included. Include information that will assist potential lenders in understanding your projections. Lenders will give more credence to the assumptions your projections are based on as they do the numbers themselves.

Business Plan Pitfalls

The following are some of the most significant pitfalls that most executives and master planners face in building and maintaining their business plans. While they are not inclusive, they represent a consensus of belief on this subject area. (The readers are encouraged to add their own pitfalls as they see fit.)

- *Focusing too much on the future.* It is commonly accepted that investment potentials and valuations for organizations are based on an organization's projected future performance. However, since the best indicator of future performance is past performance, or a company's past track record, business plans need to clearly show what milestones and accomplishments the organization has already achieved. A prior success in achieving goals tends to give investors the confidence that the management team will be able to deliver in the future.
- *Making financial projections too aggressive.* Since many investors go directly to the financial section of the business plan, it is critical that the assumptions and projections be realistic. Plans that show penetration, operating margin, and revenues per employee figures that are poorly thought-out and reasoned, or are internally inconsistent or simply unrealistic, greatly damage the credibility of the entire business plan. In contrast, clear, well-reasoned financial assumptions and projections communicate operational maturity and credibility. By accessing and basing projections on the financial performance of public companies in their marketplace, ventures can demonstrate that their assumptions and projections are attainable.
- *Not including successful companies in the competitive discussion.* Too many business plans try to show how unique their organization is and do not demonstrate their knowledge of the competition, which usually has a negative connotation. If few organizations are in a marketplace, the implication is that there may not be a large enough customer need to support the venture's products and/or services. When positioned properly, including successful or public companies in a

competitive space can be a positive sign, as it implies that the market size is big. It also gives investors some assurance that if the management team executes well, the organization has substantial upside profit and liquidity potential.

- *Asking investors to sign a nondisclosure.* Many investors will not want to sign a nondisclosure agreement because a business' strategy and concept are typically not confidential.

EXAMPLES

These were discussed in the section of this chapter entitled How to Use the Tool.

SOFTWARE

Some commercial software available includes but is not limited to

- MindMap: www.novamind.com/
- Smartdraw: www.smartdraw.com/
- QI macros: http://www.qimacros.com

SUGGESTED ADDITIONAL READING

Asaka, T. and Ozeki, K., eds. *Handbook of Quality Tools: The Japanese Approach*. Portland, OR: Productivity Press, 1998.

Brassard, M. *The Memory Jogger Plus*. Milwaukee, WI: ASQ Quality Press, 1989.

Harrington, H.J. and Lomax, K. *Performance Improvement Methods*. New York: McGraw-Hill, 2000.

Nanus, B. and Dobbs, S.M. *Leaders Who Make A Difference: Essential Strategies for Meeting the Nonprofit Challenge*. San Francisco: Jossey-Bass, 1999, pp. 279–426.

5

Comparative Analysis

S. Ali Bokhari

CONTENTS

DEFINITION

Comparative analysis is the item-by-item detailed comparison of two or more comparable alternatives or processes or products or sets of data or systems.

USER

This tool can be used by individuals, but its best use is with a group of four to six people. Cross-functional teams usually yield the best results from this activity.

OFTEN USED IN THE FOLLOWING PHASES OF THE INNOVATIVE PROCESS

The following are the seven phases of the innovative cycle. An X after the phase name indicates that the tool/methodology is used during that specific phase.

- Creation phase X
- Value proposition phase X
- Resourcing phase
- Documentation phase
- Production phase
- Sales/delivery phase X
- Performance analysis phase

TOOL ACTIVITY BY PHASE

- During the creative and value proposition phases, the concept is compared to the competitors' products or services to make a business decision if the concept is going to be value added to the organization. Projects are frequently dropped at this point when the comparison is not favorable to the organization.
- During the sales and marketing phase, both sales price and marketing strategy is compared to the competition and effort to maximize the value to the organization.

HOW TO USE THE TOOL

Whereas comparative analysis is used as a simple tool of innovation by professionals worldwide, it is also famous as a robust methodology of academic research. As our objective here is to understand comparative analysis and its use as a simple tool of innovation, the scope of our definition of comparative analysis and examples will, therefore, primarily be focused on our objective. An advanced understanding of comparative analysis methodology, however, will be helpful in gaining its in-depth knowledge.

According to the business dictionary (www.businessdictionary.com), comparative analysis is "the item-by-item comparison of two or more comparable alternatives, processes, products, qualifications, sets of data, systems, or the like." For example, in accounting, changes in a financial statement's items over several accounting periods are normally presented together to identify the emerging trends in the company's operations and financial results.

Comparative analysis as a data analysis technique is also very popular throughout the world especially in social sciences. It is used for analyzing data sets by listing and counting all the combinations of variables observed in the data set, and then applying the rules of logical inference to determine which descriptive inferences or implications the data supports.

Why Comparative Analysis Is a Useful Tool or Methodology

When we undertake comparison of two or more comparable alternatives, processes, products, or systems, our natural expectation is to find out something new (similarities and differences) that is unknown to us. For example, if we want to improve our customer service, we can try to compare it, on the selected parameters, with that of the best-known customer service providers in our own industry or even outside of our industry. In this way, comparison by its very nature leads us toward synthesis of a new knowledge—construction—which forms the basis of innovation.

Comparison is a fundamental tool of analysis. It sharpens our power of description, and plays a central role in concept-formation by bringing into focus suggestive similarities and contrasts among cases. Comparison is routinely used in testing hypotheses, and it can contribute to the inductive discovery of new hypotheses and to theory-building.

Collier 1993, *The Comparative Method*, p. 105

Comparative analysis is an effective tool that can be used as a methodology of innovation. It involves the following main steps:

- Step 1—Define the objective(s) of your comparative analysis; for example, to identify a trend; to choose the best available alternatives; and to improve a process, product, system, outcome, or result.
- Step 2—Choose a number of parameters for comparison of two or more comparable alternatives, processes, products, systems, etc.
- Step 3—Select different comparable cases.
- Step 4—Obtain and fill in data from the selected cases against each parameter.
- Step 5—Draw findings (similarities and differences) by comparative analysis and prepare a statement of your findings, or develop a strategy or action plan to achieve the objective(s) of your comparative analysis.

Advantages

1. The major advantage of comparative analysis is that it helps in the generation of new knowledge, which forms the basis of innovation. Comparison of two or more comparable alternatives, processes, products, or systems against the chosen parameters brings out similarities and differences that, in turn, lead to synthesis of new knowledge that was unknown before the comparison.
2. Comparative analysis helps in identifying trends and problems as well as in understanding problem(s), thus leading to the development of suitable solution(s) to the identified problem(s).
3. Qualitative comparative analysis brings out rich data that helps in understanding complex phenomena.

Disadvantages

1. Selection of parameters and comparable cases to conduct comparative analysis and find similarities and differences requires a good level of knowledge and skills in using this technique.
2. An inappropriate choice of parameters and cases would result into misleading findings.

SUMMARY

Comparative analysis is a very useful tool or methodology for innovation. Two or more comparable alternatives, processes, products, systems, or cases are compared against selected parameters to identify similarities and differences of the cases. This naturally leads to synthesis of new knowledge that was previously unknown. The new knowledge as a result of comparative analysis thus provides a foundation for innovation. Although there are great advantages of comparative analysis as a tool of innovation, special care in selecting parameters of comparison is needed.

EXAMPLE

Simple Example

If you want to identify the trend of your sales growth for the last two years, a simple comparative analysis of your last two years' annual sales can be performed as shown in Table 5.1. Here, your annual sales represent the parameter and for comparison you can use the actual annual sales figures of say 10M and 9M, respectively, for the last two years. This simple comparison is showing that your sales growth is 1M less or 10% negative as compared to the last year's sales.

TABLE 5.1

Comparative Analysis of the Last Two Years' Annual Sales to Identify the Growth Trend

Parameter	Year 2012–2013	Year 2013–2014	Annual Sales Growth
Annual sales	$10,000,000	$9,000,000	−10%

Complex Example

Problem/Objective: You want to improve your present customer retention rate.

You would carry out the following steps:

- Step 1. Define the objective: improve your customer retention rates by comparing your customer retention process with the similar process at other organization(s) having better and positively growing customer retention rates.
- Step 2. Choose (research) a number of parameters (such as employees' training and behavior at the workplace, product or service value, customer impression via complaints or recommendations, etc.) for this comparative analysis.
- Step 3. Select two or more comparable cases that perform better in customer retention.
- Step 4. Obtain data from the selected case(s) to find how they are performing on each of your chosen parameters.
- Step 5. Compare data (yours and others) on each of your chosen parameters and find similarities and differences, and prepare an action plan to improve your own process for better customer retention rates.

From the above example, it is obvious that on the basis of findings of the comparative analysis, you can develop an action plan to improve your customer retention rate.

Altron Group of New York has planned to set up a manufacturing plant for providing agro equipment, refrigeration, services, and industry machinery to farmers. This plant is planned to be located in a rural area with an expected investment of $40 million and generation of 750 full-time employment opportunities. The chief executive officer (CEO) of Altron has already identified seven different states, namely, Alabama, Georgia, North Carolina, Ohio, South Carolina, Tennessee, and Virginia. This plant could be potentially located in any one of these states, as each of the identified states offer a special *incentives package* to attract new investment in the state. The CEO has recently directed the chief financial officer (CFO) to conduct a comparative analysis of the incentives package offered by each of the selected states so that the best possible decision can be made with regard to the location of the proposed plant. The CFO carefully chooses the parameters of comparison and gathers all the necessary data from each of the selected states and prepares the data for comparative analysis (see Table 5.2).

Altron Group's Rural Project Profile

Location: Rural
SIC: 358 (Refrigeration and service industry machinery)
Investment: $40 million
Jobs Created: 750

From the comparative analysis of the incentives package offered by each of the seven identified states, as shown in Table 5.2, it was found out that Alabama could offer the best package if the proposed rural project of Altron Group could be located there. Thus, on the basis of this comparative analysis of the incentives package offered by each of the seven states, Altron Group decided to locate their rural project in the state of Alabama. (Case Study Data Source: http://www.hks.harvard.edu/case/ncbattle/briefing/incentiv /incpack.htm#anchor1255971, accessed: December 10, 2014.)

SOFTWARE

- Tableau Desktop Software: This is an easy-to-use and award-winning software for comparative analysis with multiple options and colored graphics. The software website provides a detailed training video with different cases through colorful and graphical illustrations: http://www.tableausoftware.com/products/desktop /download-now?ref=lp&signin=84e72805f7c9033a4802efec17862 cfc&1[os]=windows.
- fsCQA 2.5: The most popular software for qualitative comparative analysis. This software accompanies a detailed user's guide and an installation manual. This software can be downloaded from http:// www.socsci.uci.edu/~cragin/fsQCA/download/Setup_fsQCA.exe.

TABLE 5.2

Estimated Incentive Packages by State

General Incentive Category	Alabama	Georgia	North Carolina	Ohio	South Carolina	Tennessee	Virginia
Jobs tax credit	$0	$11,250,000	$2,100,000	$6,180,000	$3,750,000	$2,250,000	$700,000
Enterprise zone credits	$1,875,000	$0	$0	$750,000	$0	$0	$4,154,764
Site preparation grant	$400,000	$200,000	$250,000	$0	$0	$1,000,000	$300,000
CDBG grants	$200,000	$0	$0	$0	$0	$0	$700,000
Investment tax credit	$0	$0	$0	$0	$0	$0	$0
Other	$10,000,000	$0	$0	$0	$600,000	$0	$0
Total	$12,475,000	$11,450,000	$2,350,000	$6,930,000	$4,350,000	$3,250,000	$5,854,764

SUGGESTED ADDITIONAL READING

Collier, D. The comparative method. In: A.W. Finifter, ed. *Political Sciences: The State of the Discipline II*. Washington, DC: American Science Association. pp. 105–119, 1993.

Houlihan, B. *Sport, Policy, and Politics: A Comparative Analysis*. London: Routledge Publications, 1997.

Legewie, N. *An Introduction to Applied Data Analysis with QCA*. Available at http://nbn -resolving.de/urn:nbn:de:0114-fqs1303154, 2013.

Nelson, R.R. *National Innovation Systems: A Comparative Analysis*. Oxford, UK: Oxford University Press, 1993.

Pollitt, C. and Bouckaret, G. *Public Management Reform: A Comparative Analysis*. Oxford, UK: Oxford University Press, 2000.

Ragin, C. *The Comparative Method: Moving Beyond Qualitative and Quantitative Strategies*. Berkeley, CA: University of California Press, 1987.

6

Competitive Analysis

Achmad Rundi

CONTENTS

DEFINITION

Competitive analysis consists of a detailed study of an organization's competitor products, services, and methods to identify their strengths and weaknesses.

OFTEN USED IN THE FOLLOWING PHASES OF THE INNOVATIVE PROCESS

The following are the seven phases of the innovative cycle. An X after the phase name indicates that the tool/methodology is used during that specific phase.

- Creation phase X
- Value proposition phase
- Resourcing phase
- Documentation phase
- Production phase
- Sales/delivery phase X
- Performance analysis phase

TOOL ACTIVITY BY PHASE

- Creation phase—The information from competitive analysis is used during the creation phase to validate what needs are not being met by competitors based on their segments and product level details, and to adjust our marketing effort once we launch the innovative product.
- Sales/delivery phase—The information collected related to the competitions' strengths and weaknesses is used to refine the marketing and sales strategy.

HOW TO USE THE TOOL

Competitive analysis helps us understand the competitors that we are competing with, and/or will be competing against, by identifying the competitors, knowing these competitors in detail, and anticipating their move or response to our own competing strategy, and use the information to our advantage (Beard, 2013).

We need to understand the competing powers in addition to understanding the industry we are in and the ability that we have in developing our products or services. We use PEST and Porter's Five Forces Analysis to assess the macro economy and industry where our product will compete. On the other hand, SWOT analysis can be used to gain understanding from our product's capacity perspective, whereas competitive analysis can help in understanding from the competitors' perspective.

Definition: *PEST analysis* (political, economic, social, and technological analysis) describes a framework of macro-environmental factors used in the *environmental scanning* component of *strategic management*. It is a part of the external analysis when conducting a strategic analysis or doing *market research*, and gives an overview of the different macro-environmental factors that the company has to take into consideration.

Definition: *Porter's five forces analysis* is a framework to analyze the level of competition within an industry and the business strategy development. It draws on industrial organization economics to derive five forces that determine the competitive intensity, and therefore the attractiveness, of an industry.

The objective of competitive analysis is to assess the competitors so we can present the best competing strategy, whether to pin our product "against the competitors' at our advantage or to compete without posing threats to competitors by being in their shoes so we know enough how they think and behave" (Czeipiel and Kerin, 2011). In practicality, we have to be able to understand the situation as the competitors see it.

Competitive analysis will focus on two main areas. The first one is to get the information on important competitors. The second one is to leverage the information to understand the competitors' behavior. To do so, there are different processes, but they have a common theme as mentioned above. The following steps are based on the work of New York University Stern's Marketing Professor John A. Czeipiel and Edwin Cox School of Business Marketing Professor Roger A. Kerin. The article describes the following processes to complete competitive analysis (Czeipiel and Kerin, 2011):

I. Identifying the competitors
   When we want to identify the competitors, we want to identify them from two sides. The first one is the demand-side base where we want to identify the competitors that address similar consumers' set of needs. The second one is the supply-side base where we want to identify the competitors who use the same resources to achieve our focal business' objective. These resources can be technology, labors, operations such as supply chains, and the like.
   a. Three domains to identify whether the competitors are direct or less direct, using the same resources or not:

      i. *Area of influence*: We want to know the territory, market, business, or industry where the firms are competing to serve the same consumers.

      ii. *Immediately contiguous area*: This domain is where firms compete indirectly for the same consumers but with different resources. For example, Pepsi and Tropicana compete for the same consumer who wants to consume beverages.

      iii. *Area of interest*: This involves firms that have the same resources but do not address the same consumers; however, they have the capability to satisfy similar needs.

  b. Different level of competing firms' identification:

      i. *Product level*: Examples are functions, technology used, and materials. These components can also help identify potential competitors who use different resources in technology, market, and access that eventually serve the same consumers.

      ii. *Firm level*: Multimarket and other products from different geographies that are parts of the competitor's overall strategy to win the consumer segment in an industry that can influence our focal business action.

  c. Competitive blind spot:

    Please be aware that there are blind spots that lead to inaccurate identification of competitors. This can cause overlaps that do not list the competitors to be considered. We need to take into consideration the following to minimize the effect of blind spots:

    – Misjudging industry boundaries
    – Poor identification of competitors
    – Overemphasis on competitors' visible competence
    – Emphasis on where, not how, to compete
    – Faulty assumption about competitors
    – Paralysis by analysis

II. Identifying competitors' information needs

When our focal business (primary product or services that the organization was founded to provide) is able to implement our strategy and be able to anticipate future competitors' response against our strategy, we have achieved the goal of the competitors' analysis. To do so, we need information that is both quantitative and factual, and also qualitative and intentional. The first type will help us understand what the competitors are doing and can do. The latter type of information will help our focal business predict what the competitors are likely to do. The four key pieces of information to develop this knowledge are as follows:

1. The competitors' marketplace strategy consisting of scope, posture, and goal
2. The sources of their competitive advantage
3. The interpretation of signals sent by the competitors
4. A competitive response profile
   a. Competitors' marketplace strategy
      The marketplace strategy components will be
      - *Scope*: This defines the products the competitors offer to the market and the consumers of these products. Analysis of these two components will provide us static and dynamic information about the competitors. The static information will show us where the competitors are and what they are doing based on their current products. The dynamic part is to know where the competitors will be heading, as signaled by their choice of their next products and consumers.
      - *Posture*: This defines how each competitor acts on the market to win consumers. They usually use different modes at the same time, such as a combination from these modes: price, feature, product line breadth and depth, service level, availability, image and reputation, selling, and relationship.
      - *Goal*: This is why they are in the market. The goal is the end, whereas scope and posture are the means to get to the end. For example, Apple's intent and vision is to integrate voice, data, and video.
   b. Competitors' source of competitive advantage
      Rational competitors will take economic principles into consideration when shaping their functions and activities into a competitive advantage in a competitive market. The structure of the market can profoundly influence the conduct and financial performance of its firms (Besanko et al., 2010). Here, we need to know what are making these competitors strong and how they handle their weaknesses.

      The following are the focus when understanding the competitors' advantage:

- *Input*: We start with the inputs that feed into their value chains of production. Competitors will be selective in their inputs. Not all competitors have the luxury to have complete vertical integration. Thus, assessing their cost of the only added-value raw material is an important analysis. Information on inputs can be about the suppliers, transportation cost, then labors, and the weighted average cost of capital (WACC). The lower the WACC, the lower the hurdle rate in order to get investors' approval and expand faster.
- *Technology*: Understand the current and future direction of the technology used. For observing the competitors' signal for use of technology, published patents are some sources.
- *Operations*: We need to know the value provided from the operation. Is it effective? We can reverse engineer the product. We also can experience the quality and operations when we pay for the competitors' service, for example. The cost of goods sold is also another way.
- *Products*: Assess the products as the consumers see them. Consumer survey is a tool. The survey will show what the consumers value, and we need to assess what part of the product quality addresses that value.
- *Access, segment, and customers*: We need to know the segment it is covering, and the channel they use and the quality of the channels to reach the segments that position the competitors with advantage.

Assessing and Interpreting Competitive Signals and Actions

As we lay down the foundation of competitive analysis with identification and information, we need to build on it. The power of the information that we have on who are the competitors and what information we have on each of the competitors is that we can understand their moves in a timely manner and respond accordingly. For example, "it often means having only an hour or two to interpret the meaning of a competitor's 10% across-the-board price cut and to formulate a response" (Czeipiel and Kerin, 2011). Another situation is that we are ready for the reactions of the competitors to our product announcement. These competitors will send out signals. We need to understand the forms that each competitor uses to send the signals. We also need to understand the underlying purpose. Other domains to understand or capture this signal are the veracity (truth, bluff, misleading), forum or medium (where they send the signal), and the content (goal, internal situation, expectation of competitors' behavior, or rule of game) of the message (Czeipiel and Kerin, 2011).

We also need to understand the nature of their competitive responses or action toward the product in the market or our focal business. The first step is the type of action, whether it is frontal, where it positions its product directly to our product; flanking, where it positions itself on adjacent product–consumer segments instead of head to head; or no action at all.

Response Profile

"Analysts can take some more concrete steps as shown in Figure 6.1, which portrays a helpful framework for analyzing present and potential competitor moves and responses. The combination of this analysis of competitors' goals and assumptions together with competitors' current strategies and capabilities allow one to estimate their response profiles. A response profile tells one what kinds of actions a competitor is likely to take, if any,

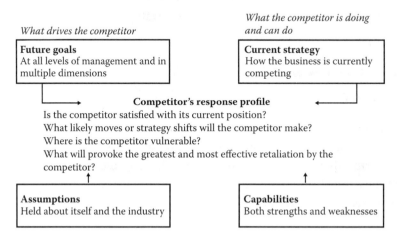

FIGURE 6.1
Competitor's response profile framework. (From Porter, M. *Competitive Advantage*. New York: The Free Press, 1980, p. 49.)

in response to the firm's own actions. Again, what this means is that you have to be able to think like your competitor."

The important keywords are the competitor's *reactiveness* and *relative clout*. Reactiveness is basically the incentive to the competitors to react to our focal business. We can measure the incentive by estimating the contribution or loss, such as revenue, profit, or brand exposure, that the competitors will have by reacting or not reacting to our focal business move. The relative clout is the likelihood of response based on competitors' size, cash position, distribution coverage, channel ownership, and relative number of salespeople.

"It may sound simplistic, but one of the most powerful determinants of a competitor's future actions is the set of economic outcomes that would result from each different competitive response. To the extent, then, that one can calculate the financial results that would flow from different actions, one should be able to predict competitors' actions." When competitors make response measurable by economic principles, we can estimate what actions they will take toward our product just so they can achieve or sustain their future economic outcomes. For example, on the basis of their response profile that we compile, if they do not see our product or service as a threat to their future economic outcomes, the nature of action will be no action at all. Another possibility is that their action can threaten our competitive edge. The output of this competitive analysis can be valuable when we try to complete another analysis (e.g., Porter's Five Forces Analysis) or feed into our business strategy.

EXAMPLE

The following is an organization's software flowcharting output and copies of four of its competitions' flowcharting software package outputs. By comparing them, the programmers can identify competitive advantages that the competition has over your approach and make appropriate adjustments (see Figures 6.2 through 6.6).

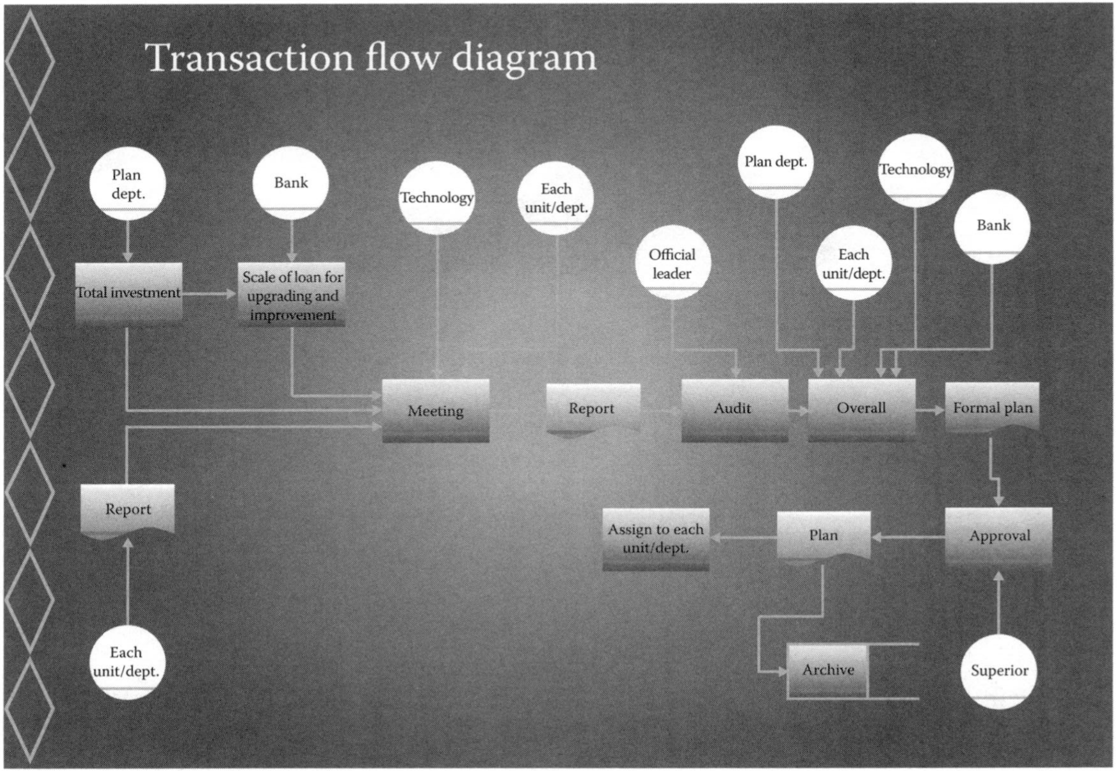

FIGURE 6.2
Comparing them for the programmers to identify competitive advantages.

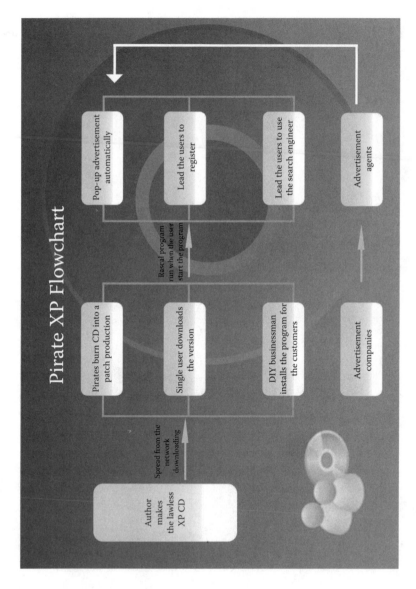

FIGURE 6.3
XP flowchart.

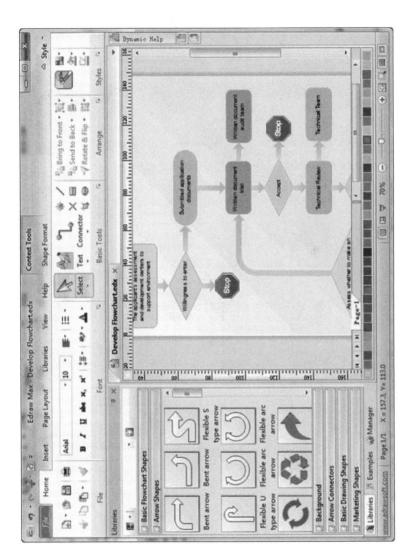

FIGURE 6.4
Another type of flowchart.

Procedure flowchart sample

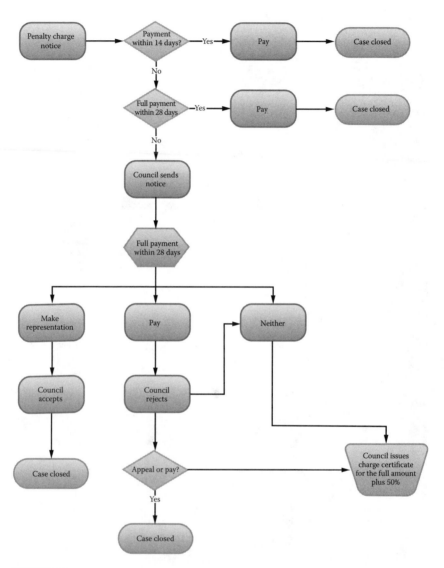

FIGURE 6.5
Flowchart collection process.

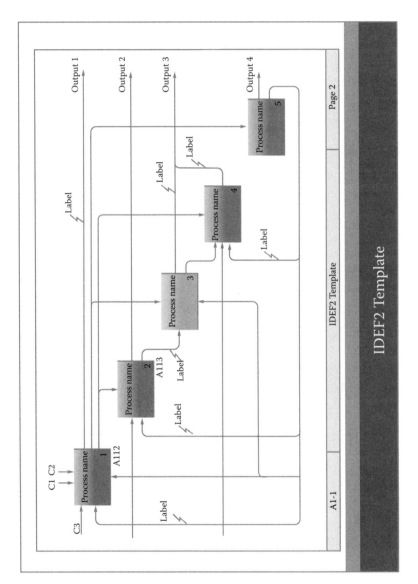

FIGURE 6.6
IDEF2 flowchart template.

SOFTWARE

- Brandwatch.com: for monitoring competitors' signals on multiple online forums/medium
- Pricemanager.com: for monitoring price elasticity as response to competition
- Blekko.com: a consolidated search for each competitor on price, technology used, consumer reviews, and company profile

REFERENCES

Beard, R. *Competitor Analysis Template: 12 Ways to Predict Your Competitors' Behaviors.* Retrieved from Client Heartbeat: http://blog.clientheartbeat.com/competitor-analysis -template/, 2013.

Besanko, D., Dranove, D., Shanley, M., and Schaefer, M. *Economics of Strategy.* New York: John Wiley & Sons, 2010.

Czeipiel, J.A. and Kerin, R.A. *Competitor Analysis.* Retrieved from pages.stern.nyu.edu: http://pages.stern.nyu.edu/~jczepiel/Publications/CompetitorAnalysis.pdf, 2011.

7

Competitive Shopping

H. James Harrington

CONTENTS

DEFINITION

Competitive shopping, sometimes called mystery shopping, is the use of an individual or a group of individuals that goes to a competitor's facilities or directly interacts with the competitor's facilities to collect information related to how the competitor's processes, services, or products are interfacing with the external customer. Data is collected related to key external customer impact areas and compared to the way the organization is operating in those areas.

USER

Usually, a team is put together to prepare a checklist of information that the competitor shopper will need to collect data on, and they will prepare a list of competitor locations that define where and how the data will be collected. Usually, the actual contact with the competitor is done by an individual. The results from the data collected are then analyzed by a team of individuals who are responsible for identifying opportunities for improving the organization's processes/services/products.

OFTEN USED IN THE FOLLOWING PHASES OF THE INNOVATIVE PROCESS

The following are the seven phases of the innovative cycle. An X after the phase name indicates that the tool/methodology is used during that specific phase.

- Creation phase X
- Value proposition phase
- Resourcing phase
- Documentation phase
- Production phase
- Sales/delivery phase X
- Performance analysis phase

TOOL ACTIVITY BY PHASE

- Creation phase—Competitive shopping activities usually take place in part of the development phase of the new product, but its information is used also in the value proposition phase and the production phase.
- Sales/delivery phase—The data collected during the competitive shopping cycle is often used to set the needs for new innovative marketing approaches and for setting future prices for the products and services.

HOW TO USE THE TOOL

It is just as important to understand the difference between competitive service levels and your service level as it is to understand the difference in product performance. Your competitors will probably be reluctant, if not hostile, benchmarking partners. The answer in the service is to buy the competitor's service. This is called competitive shopping and is used throughout the world. Safeway markets have a list of products that they can price out against other supermarkets to determine cost differentials. Department managers at The Emporium price out name brands at Macy's to be sure that they are not being undersold and that they are getting the best price from their suppliers. Hotels, banks, and airlines buy competitive services, and have their people stay in competitors' hotels, open accounts at competitors' banks, and ride on competitors' airlines. These competitive shoppers measure the competitors' performance against their own organization's performance. Also in Japan, bank employees regularly use competitive bank services to compare their bank's performance against the competitive banks' standards. Typical points to be evaluated include

- Wait time
- Ease of use
- Cleanliness
- Customer interface process
- Noise level
- Cost accuracy
- Employee capability and knowledge
- Reliability
- Predictability

In the service sector, competitive shopping is required to define the level of performance needed to compete in today's fierce market and to determine where improvement opportunities exist. It is a type of benchmarking that every service organization must repeat almost monthly.

Although many organizations develop their own staff of competitive benchmarkers, there are organizations that specialize in this activity. Evaluation Systems for Personnel (ESP) is one of these organizations. ESP provides internal and external competitive shopping data to their customers throughout the United States. Its competitive benchmarkers are known as mystery shoppers. Gerry Blumenthal, CEO of ESP, defines the organization's activities as follows:

> A mystery shopper is an individual who enters a place of business, posing as a regular customer, whose sole aim is to evaluate the level of service or sales ability. We at ESP have developed mystery-shopping questionnaires for almost every industry. Each question in our questionnaires is scored and weighted according to its specific importance. This enables us to evaluate, quantitatively, the service rendered by that specific industry or its salesperson. Thus we are able to measure performances between departments or individuals comprising a company or between different companies. This provides a tool for companies to determine where they stand in relation to the competition. The competition, if they excel in the industry, can then be mystery-shopped to "collect or learn their secrets." In effect, then, the competition has become a benchmark.
>
> During the past 8 years we have, through the eyes and ears of over 8000 mystery shoppers, established the specific "things" that people in service or sales do or do not do, in their interaction with customers or clients. Naturally, we come into contact with all levels of operators, including the most successful operators.
>
> By examining these successful operators, we have established the benchmarks for each industry. Many companies are now actively shopping the competition, specifically for this reason.

Data Analysis

Data analysis is a critical phase of the benchmarking process, because you must organize masses of numbers and statements into coherent, usable information to direct all your future activities. The success or failure of the benchmarking process depends on how well the reams of collected data are translated into usable information.

The measurement data provide you with indicators of where the best practices, procedures, and processes can be found. As you compare the data on benchmarking items against your time, you may find that you are the best (world class), the same, or worse. If you are the best, then congratulations. If your comparison is negative or the same, an opportunity exists to improve by studying another organization's or location's item.

Two types of data collected and used in the benchmarking process include qualitative data (word descriptions) and quantitative data (numbers, ratios, and so on). There has been much debate over which to collect first, and how to use each type of data. In reality, your benchmarking strategy should be designed to collect both types of data to identify potential opportunities.

EXAMPLES

Sample Data Collected Friday Night at 6 PM

	A+ Hotel	**Our Hotel**
Wait time to get registered	8 minutes	10 minutes
Ease of use	Just asked for my name	Asked for my name and credit card
Cleanliness	8	8
Customer interface process	Business like	Friendly
Noise level	9	5
Cost accuracy	$105	$120
Employee capability and knowledge	Good	Excellent

SOFTWARE

None listed.

SUGGESTED ADDITIONAL READING

Harrington, H.J. *The Complete Benchmarking Implementation Guide.* New York: McGraw-Hill, 1996.

Harrington, H.J. and Harrington, J.S. *High Performance Benchmarking—20 Steps to Success.* New York: McGraw-Hill, 1996.

8

Contingency Planning

S. Ali Bokhari

CONTENTS

DEFINITION

Contingency planning is a process (WhatIs.com) that primarily delivers a risk management strategy for a business to deal with the unexpected events effectively and the strategy for the business recovery to the normal position. The output of this process is called *contingency plan* or *business continuity and recovery plan* (Study.com).

USER

A cross-functional team of four to eight people led by a senior management leader and facilitated by a contingency planning professional gives the best results.

OFTEN USED IN THE FOLLOWING PHASES OF THE INNOVATIVE PROCESS

The following are the seven phases of the innovative cycle. An X after the phase name indicates that the tool/methodology is used during that specific phase.

- Creation phase X
- Value proposition phase X
- Resourcing phase X
- Documentation phase X
- Production phase X
- Sales/delivery phase X
- Performance analysis phase X

TOOL ACTIVITY BY PHASE

Undesirable changes or unexpected problems can occur during all phases of the innovation cycle. In each phase, the innovator should assess the activities and define potential negative things that could occur. Once the potential negatives have been identified, contingency planning should be developed to minimize the time to react to the negative shifts in a minimal amount of time.

HOW THE TOOL IS USED

Contingency planning is considered a tool of innovation because it provides an innovative roadmap or strategy (solution) to respond coherently and effectively to unexpected and disastrous events (such as heavy snowfall in the winter, fire in the workplace, earthquakes, flooding, terrorist attack, or hard drive crashes), for the full recovery of the business to normal.

To draw up a contingency plan, the following steps are recommended.

Step 1. Recognize the Need for Contingency Planning

The first step to draw up a contingency plan is the organizational recognition of the need for contingency planning because it requires strong commitment of senior management, allocation of necessary resources, and involvement of people in the organization. If any of these factors is missing, there cannot be a well-thought-out and well-practiced effective contingency plan in place—a plan that will not only help the organization to deal effectively with any unexpected eventuality but also provide with a clear roadmap of the business recovery to the normal position. Seese and Wiefling (2010) have rightly observed: "It (contingency plan) is a major undertaking requiring senior management support, end-user buy-in, and the cooperation of representatives from every area in your organization. Anyone who has tried to build consensus on something simple knows that this kind of collaboration doesn't come easily!" In our present difficult times, not having an adequate contingency plan could be completely disastrous for an organization. All well-managed businesses today is expected to have a well-thought-out and a well-practiced contingency plan in place.

Step 2. Identify Possible Contingencies

The second step to develop a contingency plan is identification of possible contingencies that an organization may have to deal with. It could be various things: (a) natural disasters like heavy snowfall in the winter, flooding, earthquake, and so on; (b) man-made disasters like terrorist acts; (c) system disruptions like hard drive or server crashes; and (d) business issues like competitors' hostile move, incapacity to fulfill commitments, declining customers, negative cash flows, rising and uncontrollable debt, sustainability of operations, acquisition and retention of necessary talent, and so on.

Step 3. Determine the Likelihood of Specific Contingencies and Assess Business Impact

The third step is to determine the likelihood, for each specific contingency identified in step 2, and consequences, assessing associated business impact or risks in terms of loss of assets (both tangible and intangible), people, processes, and customers. For example, in the United Kingdom, falling of heavy snow in the winter could be a specific contingency a business may have to face. In this case, one of the consequences would include the inability of the employees to reach the business premises. Similarly, there could be other severe to insignificant consequences of this event. To determine the likelihood (low, medium, high) of an eventuality (specific contingency identified in step 2 above) and its consequences or impact (low, medium, high) for a business, the contingency likelihood and business impact matrix in Figure 8.1 may be used.

Step 4. Determine the Strategy to Deal with the Likely Business Impact

The fourth step in drawing a contingency plan is to determine the strategy to deal with the likely consequences or business impact of the identified contingencies mitigating associated risks, and ensure necessary provisions for continuation of the business and its full recovery back to normal. Various mitigation options such as risk avoidance, risk reduction, and risk transfer are employed depending on the nature and severity level of the consequences.

Step 5. Prepare an Action Plan Identifying Responsibilities

The fifth step is to develop an action plan to address the likely consequences or business impact of the identified contingencies to ensure the

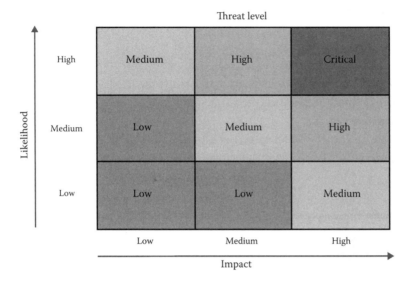

FIGURE 8.1
Contingency likelihood and business impact matrix.

continuation of the business and its full recovery back to the normal position within the estimated time frame. In addition to allocation of the required resources, the people's roles and responsibilities are clearly identified in the action plan for the effective implementation of the action plan. Necessary training of the people and allocation of the necessary resources are duly provided in the action plan.

Step 6. Execute and Test the Plan—Crisis Simulation

The sixth step is to execute and test the contingency plan so that the effectiveness of the plan is assessed through crisis simulation. This ensures widespread communication of the contingency plan for the people's awareness, and provides training and practice opportunities for the people.

Step 7. Review the Plan for Continuous Improvement

The last step is to review the contingency plan by the management, as a part of an annual strategic planning review cycle, in the light of the execution and testing of the plan for continuous improvement of the plan.

EXAMPLE

GST Group of Aberdeen developed a contingency plan identifying a number of contingencies. Two of the contingencies identified included (a) heavy snow preventing key staff attending work and (b) head office damaged by fire.

The Contingency Plan (2014) of the GST Group addressed the above-mentioned two contingencies as given in Table 8.1.

BENEFITS AND ADVANTAGES OF CONTINGENCY PLANNING

1. Keeps business running during crisis
2. Avoids panic at the time of crisis
3. Keeps losses to a minimum
4. Generates and maintains customer interest
5. Attracts investors' interest

Why Contingency Plans Usually Fail

The research shows that the following factors are the primary causes of failure of contingency plans in businesses and organizations:

1. No one is responsible for the execution of the plan.
2. The senior management is not fully committed to finance the plan.
3. The contingency plan is not a part of the annual planning review cycle.
4. There is a lack of the people awareness, training, and practice of the plan.

Point of Reflection

Can we afford to ignore contingency planning? We all, especially senior management teams, need to reflect carefully on whether we can afford to

TABLE 8.1

GST Group Contingency Plan (2014)

Contingency or Event	Likelihood of the Event	Impact on Business	Strategy to Deal with the Event	How	Who	By When
Heavy snowfall preventing key staff attending work	High	High	Key staff to work from home during the event	(a) Provide key staff with laptops and remote access to the firm's network (b) Provide key staff with mobile phones	IT department	June 30, 2014
Head office damaged by fire	Low	High	(a) Run the business from temporary premises (b) Take the firm's data with live backup to cloud computing	(a) Source firms that provide temporary business premises and agree to a provisional contract (b) Source firms that provide cloud computing facility, agree to a contract, and get the facility	Property services IT department	January 30, 2014 January 30, 2014

ignore contingency planning as a strategic top priority or as a tool of innovation that can help us better manage major disasters such as the

- Tsunami Japanese nuclear power plant disaster in 2011
- British Petroleum (BP) Mexico Gulf oil spill disaster in 2010
- Twin Towers 9/11 disaster in 2001

SOFTWARE

The following software may be used for preparing a contingency plan. Teamwork facility is available with these software. Free trial and training tutorials are available from the software vendor.

- Smartsheet: www.smartsheet.com
- Resource Planning Made Easy: http://www.retaininternational.com

When we expect rain and go out, we tend to keep with us an umbrella; this is our contingency plan.

REFERENCES

Seese, M. and Wiefling, K. *Scrappy Business Contingency Planning: How to Bullet-Proof Your Business and Laugh at Volcanoes, Tornadoes, Locust Plagues, and Hard Drive Crashes.* ISBN: 9781600051500. Cupertino, CA: Happy About, 2010.

Study.com. What is contingency planning in business?—Definition, example & importance. Available at http://education-portal.com/academy/lesson/what-is-contingency-planning-in-business-definition-example-importance.html#lesson. Accessed January 10, 2015.

WhatIs.com. Definition: Contingency plan. Available at: http://whatis.techtarget.com/definition/contingency-plan. Accessed January 10, 2015.

SUGGESTED ADDITIONAL READING

International Federation of Red Cross and Red Crescent Societies. Contingency planning guide. Available at http://www.ifrc.org/PageFiles/40825/1220900-CPG%202012-EN-LR.pdf. Accessed January 12, 2015.

9

CO-STAR Value Propositions and Pitching Tool*

Lisa Friedman and Herman Gyr

CONTENTS

---

* This chapter is a condensation from the book *Creating Value with CO-STAR: An Innovation Tool for Perfecting and Pitching Your Brilliant Idea*, by Laszlo Gyorffy and Lisa Friedman.

Innovation springs from twin sources: the profound human interest in achieving something great—a natural yearning for pioneering, for discovery, for new frontiers—coupled with systematic processes for continuously creating and improving customer value. When the ingenuity and aspiration of a group are paired with the discipline of innovation, magic happens.

Herman Gyr
Co-founder, Enterprise Development Group

DEFINITION

The CO-STAR innovation methodology is a tool and a template that helps users to

1. Turn their ideas into compelling value propositions
2. Iterate these with others to make each value proposition as strong as it can be
3. Create a powerful pitch for their value propositions to help secure the approval, resources, team members, or funding required to move forward

USERS

CO-STAR is a tool to guide innovators and their teams to turn ideas into compelling value propositions. Once users create their CO-STARs, the next step is to engage others whose information and expertise could help make these value propositions even stronger. CO-STAR is also a helpful tool for leaders and managers, investors, or judges in innovation competitions, who have to evaluate a large number of value propositions and select a few to move forward. If innovators present their ideas using a common template for their value propositions, the selection process is far easier for those who have to approve, fund, and mentor winning projects.

OFTEN USED IN THE FOLLOWING PHASES OF THE INNOVATIVE PROCESS

The following are the seven phases of the innovative cycle. An X after the phase name indicates that the tool/methodology is used during that specific phase.

- Creation phase X
- Value proposition phase X
- Resourcing phase X
- Documentation phase X
- Production phase X
- Sales/delivery phase X
- Performance analysis phase X

HOW TO USE THE TOOL

Overview

CO-STAR is specifically designed to focus the creativity of innovators on ideas that matter to customers and have relevance in the market. It is an easy-to-use tool for turning raw ideas into powerful value propositions.

CO-STAR unleashes passion for an emerging value proposition that comes from a consistent focus on *discovering and optimizing its value*—by the innovator, and by all those who engage in the exchanges and discussions for how to make it better.

Entrepreneurs, innovators, and funders alike can use CO-STAR to build and successfully pitch value propositions that are smart enough to launch market-worthy new ventures, and strong enough to make a difference.

CO-STAR helps its users to

- Identify significant new opportunities to create compelling customer value
- Capture the essence of an idea and communicate it quickly and clearly to others
- Collaborate with others to get their input and to integrate collective expertise and intelligence into the value proposition
- Craft a compelling pitch

What Is CO-STAR?

CO-STAR assists in the formation of a *value proposition*: a crisp and compelling description of how an innovator intends to bring a solution to a specific customer group in a manner that delivers more value than competing alternatives. Some CO-STAR value propositions may be used to innovate existing products, services, or business models, while others may present bold game-changing offers that disrupt and transform a marketplace.

CO-STAR is an acronym that focuses on six key elements for turning an idea into a value proposition:

Customers and their most important needs are at the heart of relevant innovation.

Opportunity exists within the range of new developments (technical, social, behavioral, competitive) that open promising new market spaces.

The C and O (Customer need and market Opportunity) exist whether the innovator does anything or not. The innovator identifies the CO and makes it visible to others. The STAR is then the innovator's response to a remarkable customer and market opportunity.

Solutions are the innovations that uniquely serve the identified Customer needs that simultaneously take advantage of the most significant Opportunities in the market.

Teams are the assembled talents for enabling the success of the Solution.

Advantage is the unique selling proposition over the available competitive alternatives.

Results are the outcomes that flow from the successful implementation of the value proposition (which include several other Rs as well—*Rewards* for customers, *Return on Investment, Revenues* and profits, as well as any high-impact *Returns* for the team, company, shareholders, environment, or society).

CO-STAR can be implemented across an enterprise using templates; books and articles (see Gyorffy and Friedman, 2012 and Suggested Additional Reading); training and coaching; as well as by using online tools and mobile apps (see Software section). This allows organizations to rapidly master value creation best practices and effectively disseminate them throughout an enterprise. The broad adoption of CO-STAR practices fosters a culture of innovation that helps establish and maintain sustainable competitive advantage that comes from the effective engagement of everyone in the enterprise.

A Typical CO-STAR Effort Begins*

Step 1. Create Initial CO-STAR

Write down your initial thoughts in the CO-STAR format (see CO-STAR template in Figure 9.1). Reflect on what you discover about your idea and decide if it is worth pursuing. If so, proceed to step 2.

Step 2. Gather Collective Intelligence

Share your initial thoughts with a few trusted colleagues. Listen and learn from their reactions and feedback. Encourage them to tell you what they like about your idea and what could make it stronger. Ask them to comment separately on each part of your CO-STAR. What do they think about the Customer, Opportunity, Solution, Team, Advantage, and Results? This feedback is *gold*—learn as much from it as you can. Do not defend your existing position—this only reduces the amount and quality of feedback you will get from people who want to help you figure out whether your idea is worth pursuing.

Once you have gathered feedback on your initial CO-STAR, you can begin to amend it to make it stronger. If you decide your value proposition has potential, proceed to step 3. If not, continue to iterate, or decide to let it go.

---

* Adapted with permission from *Creating Value with CO-STAR: An Innovation Tool for Perfecting and Pitching Your Brilliant Idea* by Laszlo Gyorffy and Lisa Friedman, published by the Enterprise Development Group Inc., 2012, pp. 29–32.

Pitching template

The hook

Begin with a compelling question, fact, or statement that generates curiosity

C   *Who is the* customer?

O   *What is the* opportunity?

S   *What is your* solution?

T   *Who needs to be on the* team?

A   *What is your competitive* advantage?

R   *What* results *will you achieve?*

The request

Conclude with a specific request regarding next steps. A meeting? Resources?

FIGURE 9.1
CO-STAR pitching template.

Potentially realizable *value* is the single most important metric for an innovator's commitment to his or her idea, or the decision to take it off the table.*

Step 3. Optimize the Value in Your CO-STAR

This pass requires research and quantification. Integrate the feedback you have already received and begin to answer the CO-STAR questions more

---

* Optimizing a CO-STAR can include using interactive software, individual assignments, virtual group work, case studies, video summaries, prototypes and pilots, personalized feedback from your CO-STAR program coach, exchanges with leaders from across the globe, and peer reviews, which are all combined to stimulate your thinking, unlock new insights, change habits, and firmly embed the learning in your specific business context.

thoroughly. Then take your idea out to a more objective audience (including possible future customers) to test its viability and potential. Repeat step 2, following a pattern of iteration of 2–3, 2–3, 2–3, until the value proposition either evolves to where it can be implemented at least at the level of a simple prototype or even a *minimum viable product,* or can be discarded because its value just cannot be proven through the various tests and inputs from others.

You Can Work CO-STAR from Two Directions

1. Start with an unmet customer need and a market opportunity and develop a viable solution.
2. Have an energizing insight and then determine if this might be relevant for an actual customer need, and whether there is a viable market for your idea.

The Internet is a remarkable vehicle for exploring and understanding the world. Use it. Use your online social network as well. And do not forget to get out of your office and look around: observe, interact, prototype, test, collect. The goal of the iteration process is to maximize potential and minimize business risk for your idea. The first draft you present should always be a CO-STAR, even if it is very cursory and simple. Then seek as much input into each CO-STAR element as you can get.

Relentless Quantification

Innovators and their supporters need to know: how many customers, how much better, how much faster, and how much cheaper. Quantification is critical to judging the potential value of an idea, calculating return, reducing risk, and limiting subjectivity.

If you begin with an idea—with a solution—it is important to go back to the beginning to validate your idea: Is there really an important customer need where your idea offers a good solution? Is there a worthwhile market opportunity? Answering these questions will slow you down from just charging ahead with your new idea, but this homework can help you "go slow to go fast." Understanding the most compelling CO can help you see opportunities for your idea that you may not have imagined before. This will help you transform and progress with your great idea into a full value proposition.

Clarifying Your Solution

For communication purposes, the description of your idea should be simple, relevant, and intuitively understandable. Use common terms to state concretely:

- What is the solution you are proposing?
- What are its key functions and features, inputs and outputs?
- Which business will you be in? What will be your business model?
- Which new technologies or intellectual property (IP) are incorporated into your solution?
- Does your solution complement or displace an existing offering?
- Which parts of the customer's total unmet need does your solution solve or not solve?
- What kinds of assets or resources are required for your solution?
- What is the expected price range of your solution for the customer?
- What is the estimated total cost of your solution?

Very often, as you complete the "S" section in your CO-STAR, you will uncover areas that need further exploration. Do not spend too much time documenting every aspect early on in your development efforts. The goal is "once over lightly." Remember that your idea will evolve over time, so there is no need to fall in love with your initial concept. You need to retain your ability to let go of parts of your idea as new and better possibilities emerge from your research and collaborations.

Managing the Paradox of Innovation

As a committed champion of your idea, you will need to manage the duality and often paradoxical nature of the innovation process—you will need to bring discipline to a creative act. CO-STAR is a thinking-and-doing tool that allows you to blend and balance the multiple demands of invention.

While everyone's journey is different, we have found over the years that you will be required to manage a number of contradictions. You will be required to

- Be passionate and committed yet willing to let go as it becomes necessary and appropriate.
- Build a vision of the future while working in the present.

- Gather broad input while staying focused on delivering specific results.
- Develop personal ideas while maintaining the perspective of the customer.
- Stay the course and stay committed while continuously seeking feedback and allowing evolution.
- Be knowledgeable and confident yet willing to admit to not knowing something.
- Be willing to take chances while working to reduce business and product risks.
- Be willing to think outside the box while working within the confines of an organization's or network's strategy and structures.
- Take personal ownership for your ideas while building a sense of shared responsibility within the team and ecosystem.

Advantages

One of the goals of CO-STAR is to bring the discipline of innovation to an organization in a practical, easy-to-use way. There is tremendous power in innovators using shared tools, as it enables innovation across a large group or ecosystem. CO-STAR works exceptionally well for this, as it is easy to use and to remember. At a time when speed to innovation is essential, simple and straightforward innovation tools provide a tremendous boost to innovators' ability to collaborate and create across boundaries.

Why CO-STAR?

Whether you are employed in a large company or a small business, doing scientific work in a laboratory, serving your country in the government, working in a nonprofit, or looking to start your own business or organization, every innovator must understand two truths:

1. The idea you generate must deliver value. Being able to crystallize your thinking and determine the true value of your idea is critical to success. Without understanding the fundamentals of value creation, you may find you are pushing an idea no one wants. Before an idea can become great, its brilliance must be understood. What value does it deliver that makes it extraordinarily attractive to customers and its potential market?

2. At some point, you will need help with your idea. No one does any kind of significant innovation on his or her own. No one has all of the necessary skills and resources. While inventors may enjoy brilliant sparks of insight, innovation is inherently a social activity. Thus, you must be able to effectively communicate your idea and collaborate with others to improve your thinking, gather support, secure funding, and put your idea into action. More often than not, ideas start in an unfinished state. Many good ideas even start out as *bad* ideas. Yet, every seemingly bad idea, when tested and iterated by a community of committed collaborators, may well become the *next big thing*.

Disadvantages

The environment in some organizations may prove to be hostile to creative thinking. Those managing the CO-STAR value creation process in these organizations are likely to encounter obstacles such as a general resistance to change, free expression being stifled by a pervasive culture of blame, or failure leading to penalties rather than to an opportunity to learn.

Also, inadequate and nonexistent incentives may lead to slow decision making, and a reluctance to think and move outside of strict job descriptions. In other enterprises, the prevalent view may be that the best ideas come from the top, along with rigid formalities and rules that often block proposals and pitches from innovators throughout the enterprise.

One of the most important requirements for the successful use of CO-STAR in a large organization is that the leaders hold a vision for the future of the enterprise that requires innovation. If leaders do not see innovation as essential to the success of their enterprise, they are not likely to encourage innovators to identify customer needs and emerging opportunities, to brainstorm ideas and turn them into value propositions, to collaborate to improve these, and to pitch CO-STARs for approval and funding.

Finally, innovators who create value propositions and want to pitch their CO-STARs need receptive *catchers*. Innovators and teams in large organizations who believe in the deep value in their CO-STARs need to be able to pitch to someone for support and resources—to leaders, partners, investors, and mentors in order to bring their solutions to life. If these leaders or a culture and innovation ecosystem do not exist, innovators and teams may simply take their CO-STARs elsewhere—by leaving the organization to found their own start-up or by taking their idea to a more receptive company or investor.

SOFTWARE

CO-STAR is currently built into two kinds of software:

- *Apps for individual innovators.* These apps enable innovators to craft their value propositions, share them with others, iterate and improve them, and ultimately pitch their value propositions for the approval, resources, staff, or funding they need. An example of this kind of app includes One-Hour Innovator, which utilizes a series of videos and prompts that guide users to create their CO-STARs (see www .onehourinnovator.com).
- *Large-scale online innovation management platforms for teams, organizations, and innovation ecosystems.* These platforms are designed for large numbers of users (from a single department, to enterprise-wide systems, to open innovation systems that include customers and users, suppliers, potential partners, and/or funders). Innovators can share their ideas and their CO-STARs with other users online, which allows them to collaborate across department boundaries, as well as across geographies and time zones.
- Current examples of online innovation management platforms that have CO-STAR built into them are from Qmarkets (www.qmarkets .net) and IdeaScale (www.ideascale.com).

These platforms provide educational information to guide users through creating their CO-STARS. This information typically offers details about challenges at hand (background information on the CO, on customer or market statistics, videos from leaders describing emerging opportunities, or any information innovators need to create effective solutions). The platforms also provide instructions and hints for creating an effective CO-STAR value proposition and guide users step-by-step through the process.

The online innovation platforms can be used for either spontaneous (bottom–up) or sponsored (top–down) innovation.

Spontaneous Innovation

Individual innovators can post ideas and CO-STARs for innovations they want to propose. The platform can help innovators find ideas that are similar to the idea or CO-STAR they are proposing, so they can choose whether to join an ongoing effort or team, or to launch a new effort. Innovators can get input from others to strengthen their ideas, and the software will often help them direct their CO-STARs to the managers or teams who could help move these forward.

Strategic Sponsored Campaigns

These online innovation platforms enable leaders to run sponsored campaigns, to issue an Invitation to Innovate for CO-STARs around a specific strategic challenge. Leaders typically give background information about a particular challenge, and they set a timeline for proposed CO-STARs to be submitted (from several days to several weeks). Some campaigns may have two or more rounds, where innovators propose initial CO-STARs, get feedback, and then enter their new, improved CO-STARs.

Online innovation campaigns typically offer innovators and teams with winning CO-STARs the promise of time and support to move forward, which can include additional resources, staff, coaching, or funding for

- Additional research needed for proof of concept
- Resources to build prototypes and run pilots
- Staff for project teams
- Connection with experts, mentors, funders, and other key stake-holders in the innovation ecosystem
- Support for new product or service launch and implementation

Using Online Innovation Platforms to Optimize the Value of CO-STARs

In gamified innovation platforms, users are encouraged to participate by earning points, badges, or other rewards for contributing their input. Users can vote for their favorite CO-STARs so that crowd favorites rise to the top. The *wisdom of the crowd* makes winning CO-STARs more broadly visible.

Users can also earn points for giving input to other users' CO-STARs, with the goal of improving these and helping innovators ensure their value propositions grow to be as strong as they can be. The CO-STAR framework helps users to give very specific and actionable feedback. For example, platform users might respond, "Your solution could also work really well with Customer X." Or, "A person who would be great for your Team is Y. She has the specific skills you said you need most."

These platforms also allow users to invite online feedback from specific users or experts in the field. Innovators can then integrate feedback and put up progressive iterations of their CO-STARs to enable them to develop and improve over time.

In general, providing broad access to an online platform for developing, sharing, and strengthening value propositions dramatically increases the quality of innovation and speeds up its development and implementation.

SUMMARY

CO-STAR is a tool and template that enables innovators to build a strong value proposition, to collaborate with others to optimize its value, to frame a compelling pitch and crystalize its value, and to communicate it quickly and effectively to others. These are critical skills every innovator needs.

One of the most important advantages of the CO-STAR model is the underlying power of a shared discipline of innovation. If organizations or innovation communities and networks have common concepts, language, tools, and practices, then innovators throughout the system can collaborate much more effectively.

In addition, using a common framework for value propositions makes the task much easier for leaders or investors who must choose a few winning value propositions from a much larger number of potential possibilities. When competing value propositions are offered using the same CO-STAR format, leaders or investors can compare which Customer groups would benefit and how significant the market Opportunities might be. They can compare Solutions across their prioritized sets of criteria. They can see how the capabilities within the Teams compare. They can judge whether teams understand their current and emerging competitive landscape, and assess which teams best frame a convincing argument that their solution offers a compelling Advantage over these alternatives. Finally, it is easier for leaders to compare the Results and Returns expected over time.

CO-STAR provides the foundation for an integrated innovation process. Because it is a tool for creating, iterating, strengthening, and pitching value propositions, using CO-STAR enables organizations to create an *end-to-end* innovation process that includes the following:

- Define a strategic opportunity or problem (leaders or innovators identify an initial CO).
- Gather trends and market insights (detailed information to clarify the O—the market Opportunity).
- Observe and gather information on Customer needs (detailed information to clarify the C).
- Generate ideas internally or externally or both (the S—Solutions).
- Develop initial ideas into full CO-STAR value propositions.
- Pitch to learn. The innovators' or teams' early pitches are to communicate their value proposition and gather collective intelligence and input.
- Capture, improve, and manage value propositions throughout a team, organization, or network.
- Build teams to move the strongest value propositions forward.
- Pitch value propositions for initial approval, funding, and support—for proof of concept.
- Evaluate and select the best.
- Prototype proposed solutions.
- Gather input and test new concepts with pilot customers.
- Integrate learning into value propositions.
- Pitch for final approval and resources.
- Innovators and teams continue to work with experts and mentors to move each value proposition forward.

Leaders in large organizations can secure valuable quick wins with *early adopters* who demonstrate to others that this process works—seeing is believing.*

Innovators typically are passionate problem solvers. They often choose to work on complex and important challenges that very often cannot be solved by one individual alone. CO-STAR is a tool that can enable innovators to grow brilliant ideas into high-impact value propositions, and to enable leaders and investors to select and fund the best.

---

* CO-STAR was developed by Lisa Friedman, Laszlo Gyorffy, and Herman Gyr based on their study of Silicon Valley value propositions and pitching practices. It is in use by enterprises in 39 countries worldwide.

EXAMPLE

Imagine you are an innovator and have one of those *light-bulb moments*. You come up with a product idea that will dramatically cut lighting costs while also being better for the planet. You know a little about the industry and are willing to make some assumptions and estimates. An early draft of your *Seeing the Light* value proposition might look like this:

Customer
- Consumers and businesses that want to reduce their electric bill through more sustainable illumination.

Opportunity
- Lighting is a $100 billion industry worldwide, where sales of traditional incandescent light bulbs have steadily declined during the last five years.
- Following Moore's law, semiconductor costs are going down continuously so the Seeing the Light *bulb* will get cheaper at a time when the world is looking for more efficient and sustainable energy solutions.
- Governments worldwide have passed measures to phase out incandescent light bulbs to promote more energy-efficient lighting. Customers have to find new lighting solutions.

Solution
- The Seeing the Light bulb is a high-performance light bulb based on light-emitting diode (LED) technology where the semiconductor made of one negatively charged layer is bonded to a positively charged layer. When electricity is introduced, electrons leave the negative layer for the positive layer to combine with atoms that have a missing electron. Light is emitted through the combination.
- We have found a unique way to manufacture this bulb at half the cost of other LEDs. Customers prefer LED light bulbs because they last longer and save energy, but they often are reluctant to buy them because the bulbs are so expensive. Our patented manufacturing process cuts the cost of production of LEDs by more than 75%, allowing them to sell them to customers at a much more attractive price, while also achieving better margins.

Team

- We have a senior team with years of experience in the lighting industry and technical expertise in LED. I will provide project management and product design expertise; Anna P. will lead the LED product development team; Jessie D. will lead the manufacturing team; and Terry K. will lead distribution and sales.

Advantage

- Over incandescent bulbs: The Seeing the Light bulb is more durable than standard incandescent bulbs, which have fragile filaments, and it lasts 100 times longer.
- The sale of incandescent bulbs higher than 40 watts will be banned by 2014.
- Over compact fluorescent lights (CFLs): The Seeing the Light bulb contains no toxic mercury, is dimmable, and lasts many times longer than CFLs.
- Over other LEDs: The Seeing the Light bulb is at least 50% cheaper than other LED bulbs for the same performance.

Results

- Without making changes to a business customer's infrastructure, the long lifespan of the Seeing the Light bulb reduces labor costs due to fewer purchases and installations.
- A family could install Seeing the Light bulbs in their newborn's bedroom and not need to replace the bulb until that child went to college. The family also would have prevented 870 tons of carbon dioxide going into the atmosphere, as the bulbs use much less electricity to produce light.

REFERENCE

Gyorffy, L. and Friedman, L. *Creating Value with CO-STAR: An Innovation Tool for Perfecting and Pitching Your Brilliant Idea.* Palo Alto, CA: Enterprise Development Group Publishers, 2012.

SUGGESTED ADDITIONAL READING

Carlson, C. and Wilmot, W. *Innovation: The Five Discipline for Creating What Customers Want.* New York: Crown Business, 2006.

Friedman, L. and Gyorffy, L. Leading innovation: Ten essential roles for harnessing the creative talent of your enterprise. In: P. Gupta and B. Trusko, eds. *Global Handbook of Innovation Science*. New York: McGraw-Hill Education, 2014.

Friedman, L. and Gyr, H. Creating your innovation blueprint: Assessing current capabilities and building a roadmap to the future. In: P. Gupta and B. Trusko, eds. *Global Handbook of Innovation Science*. New York: McGraw-Hill Education, 2014.

Friedman, L. and Gyr, H. *The Dynamic Enterprise: Tools for Turning Chaos into Strategy and Strategy into Action*. San Francisco: Jossey-Bass Business and Management Series, John Wiley, 1998.

Gyorffy, L. Direction and discipline: How leaders tap the creative talent of their enterprise. *International Journal of Innovation Science*, vol. 2, no. 2, June 2010.

Gyorffy, L. HR's role in championing innovation. *HR West*, pp. 9–12, May 2013.

Gyorffy, L. It's show time. *Leadership Excellence*, vol. 27, no. 9, pp. 11–12, September 2010.

Osterwalder, A., Pigneur, Y., Bernarda, G., and Smith, A., Bernarda, G., and Papadakos, P. *Value Proposition Design: How to Create Products and Services Customers Want*. Hoboken, NJ: John Wiley and Sons Inc., 2014.

10

Cost–Benefit Analysis

Dana J. Landry

CONTENTS

DEFINITION

- Cost–benefit analysis (CBA)—a financial analysis where the cost of providing (producing) a benefit is compared with the expected value of the benefit to the customer, stakeholder, etc.
- Confirmation bias—the tendency of people to include or exclude data that does not fit a preconceived position.

USER

A CBA can be used by an individual or a team, and is often part of a value proposition or business plan. The tool can be used by an individual but most often requires a team with sufficient knowledge of the end product, its proposed features, how customers value those features, and the current or potential cost of the component parts.

OFTEN USED IN THE FOLLOWING PHASES OF THE INNOVATIVE PROCESS

The following are the seven phases of the innovative cycle. An X after the phase name indicates that the tool/methodology is used during that specific phase.

- Creation phase
- Value proposition phase X
- Resourcing phase X
- Documentation phase
- Production phase X
- Sales/delivery phase
- Performance analysis phase

TOOL ACTIVITY BY PHASE

- Value proposition phase—The CBA is tailor made for this phase. During this phase, one or more value propositions are clearly identified. As these are identified, it is useful to quantify these benefits. In this process, the innovator might find that the benefit is not what he or she originally thought. If the data supports the low benefit valuation, then data has been collected that will feed the subsequent CBA. A quick CBA during this phase can often redirect the innovator to focus on different, a different mix, or perhaps even fewer, proposed new benefits.
- Resourcing phase—The financial output of the CBA is often a requirement for completing this phase. The innovator must be able to show that his or her selected idea is well thought out and that it justifies the amounts of financing being requested. The CBA should demonstrate the value proposition to the company as well as to the customer, and the resulting analysis from this phase is often used to review the actual outcome of the project with the forecast made in the original CBA.
- Production phase—The cost analysis in this phase is more often the straightforward type of capital investment analysis in production machinery for which many companies have their own tailor-made

spreadsheets and/or software programs. However, if the value proposition includes innovations to the production process, perhaps creating a process new to the company, then this will need to be considered in earlier phases and, if necessary, updated during this phase.

HOW THE TOOL IS USED

This method was first introduced by Jules Dupuit in the 1930s. It should not be confused with the more straightforward cost analysis where one compares only the known cost of two or more options that typically aim to accomplish the same outcome, nor with any of several other cost analysis tools. Most organizations have a standard preference for which tools are used in standard project analysis (Cellini and Kee, 2010).

CBA was used extensively in the 1950s as a way of comparing project costs versus benefits, particularly by government agencies, to determine the best way to implement a specific policy. The CBA requires that the user specify the specific benefits of the option and place a value or financial benefit to the option. There can be multiple costs and multiple benefits to any option.

In innovation, it is used to compare various ways to achieve the desired output such as the best mix of customer features most valued by the end user of a product versus what the end user is willing or able to pay. The results of a CBA are often expressed as a financial measurement such as payback period, return on investment (ROI), total sales, or similar metrics. You can use CBA in a wide variety of situations.

In a CBA, the innovator will attempt to place a value on the benefits to the customer. Innovation is very focused on finding the new, unstated, or unknown value proposition that will lead to a significant positive customer reaction. By attempting to place a value, price, or cost to these benefits, the innovator gains insight into what the customer may be willing or able to pay for the benefit, allowing the innovator to better mix the benefit–cost profile of the new product or service.

CBA is generally conducted in a series of steps. The series defined below make the assumption that the innovator has already identified multiple benefits, the cost of parts or actions needed to deliver on the various benefits, and is now at a point where he or she must start to make choices to better align the ability to deliver (the problem of limited resources) with what to deliver (you can never deliver all possible options).

1. The innovator identifies the various options by component, subsystem, system, etc.
2. List all the customers, stakeholders, etc., who will incur the cost or reap the benefit.
3. Select the metrics to be used, such as ROI, net present value, etc., that you will use. Avoid changing metrics later as it will look like you are trying to find a metric that *looks good* for your particular situation.
4. Identify the cost of the inputs to the innovation and identify the benefits that are to be converted in the same common currency used for cost.
5. Select a specific and appropriate time period for your analysis. Any financial analysis is time sensitive.
6. Tally the cost and benefits into your selected metrics. This is often done in a table or spreadsheet.
7. Perform several different scenarios, often called a sensitivity analysis.
8. Compare options and understand that this is not always looking for the biggest number but the best overall performance of the option.

By far, step 4 is where most people get into trouble. The innovator is cautioned to find the best available expertise and avoid overstating the benefits of the option. The most difficult case can be if the benefit has a direct effect on human life, as valuing a human life, while actually a tool often used by regulatory agencies, is a very difficult proposition. Innovators need to be aware of their tendency toward confirmation bias and be as honest in assigning negative values for negative benefits as they do positive for positive ones.

The second danger step is the sensitivity analysis. All too often, a particular scenario looks really good and the innovator may seek ways to justify its adoption. Remember, a really good analysis based on a very unlikely set of conditions is a tough way to build an innovation.

As mentioned, the CBA typically puts all costs into a set time frame in order to properly assess the value of the money. Innovators not schooled in finance are encouraged to put a solid finance team member in place in when performing CBAs.

Sensitivity analysis will often deal with probabilities of certain outcomes and actions as the CBA increases in complexity. Again, innovators are encouraged to add a well-trained and experienced statistician to the team during the CBA. Such a person should perform two functions. First is to make sure that all the rules of statistics and probability are followed.

Few things can be more disheartening than making a case using a CBA that is subsequently shot full of holes by a well-versed venture capitalist team member.

The second function is to convert probabilities into common language. The senior manager or venture capitalist needs to understand the risks, not the mathematical result. A statistician who can speak plainly is worth his or her weight in gold.

In the final analysis, the CBA will only be as good as the data provided and the judgments made, especially when data is scarce. Do not rely only on this one tool to make a critical selection.

EXAMPLE

In this example, we look at the potential for a new medical device. This very simplified analysis is common to the medical device industry and has the type of surprises that come with this business.

An innovator has created a concept for a new surgical device. Knowing that the industry is very cost sensitive, the innovator *is seeking to perform a CBA to further understand the benefit of the new device, the main benefits seem to be (a) less time to do the surgery and (b) easier to use.* On the basis of interviews with hospital personnel, the innovator collects the following data. All cost are a *per procedure* cost. The first thing the innovator learns is that the surgeon plays the key decision role in choosing the device, and his or her concept of benefits become key to the analysis.

	Cost	Benefit
Product cost to hospital	$500	
Profit from insurance reimbursement		$1,000
Surgeon training cost	$5,000	
Surgeon salary loss during training	$25,000	
Negative effects of initial learning[a]	$15,000	
Time saved during surgery (10 minutes)[b]		$1,000
Ease of use[b]	$0	

[a] There are such things as higher take-back rates charged to the surgeon that can occur when a new device is being used for the first few times.

[b] In this example, the innovator may think that saving a doctor 10 minutes is very valuable. He finds from the surgeons that they do not care because it is not a big enough savings to allow them to make money doing something else.

Very quickly, the innovator sees that a 10-minute reduction (easier to use benefit) does not create a compelling value proposition.

There are many examples of CBA on the web. We recommend that the potential innovator find examples that closely match the field they are working in or as close as possible. Any CBA does not have to be complex but it does have to be complete and honest.

SOFTWARE

A CBA can be easily performed on a spreadsheet, and the web offers both free and fee-based templates. The innovator is cautioned that most templates are generalized and almost always require modification, sometimes significantly so. Also, many firms have their own homegrown analysis templates. If one exists, you will almost assuredly be required to use it.

REFERENCE

Cellini, S.R. and Kee, J.E. Cost-effectiveness and cost–benefit analysis. In: J.S. Wholey, H.P. Hatry, and K.E. Newcomer, eds. *Handbook of Practical Program Evaluation*, 3rd ed., pp. 493–530. San Francisco: Jossey-Bass, 2010.

11

Financial Management/Reporting

Frank Voehl

CONTENTS

DEFINITION

Financial management—Activities and management financial programs and operations, including accounting liaison and pay services; budget preparation and execution; program, cost, and economic analysis; and nonappropriated fund oversight. It is held responsible and accountable for the ethical and intelligent use of investors' resources.

Financial reporting—Includes the main financial statements (income statement, balance sheet, statement of cash flows, statement of retained earnings, and statement of stockholders' equity) plus other financial information such as annual reports, press releases, etc.

USER

This tool can be used by both individuals and groups of any size.

OFTEN USED IN THE FOLLOWING PHASES OF THE INNOVATIVE PROCESS

The following are the seven phases of the innovative cycle. An X after the phase name indicates that the tool/methodology is used during that specific phase.

- Creation phase X
- Value proposition phase X
- Resourcing phase X
- Documentation phase X
- Production phase X
- Sales/delivery phase X
- Performance analysis phase X

TOOL ACTIVITY BY PHASE

- Creation, value creation, and resourcing phases—During these phases, the tool is used to stimulate financial thinking and to evaluate solutions to ensure that the product or process design will adequately fulfill the customers'/consumers' requirements at a reasonable cost level.

- Documentation and production phases—During these phases, financial management is used to so select suppliers and control costs to ensure costing projections are accurate.
- Sales/delivery phase—Although the selling price is driven by what the market will pay for an individual project, the sales and marketing strategy has a big impact on the quantity of sales and as such the financial return to the organization.
- Performance analysis phase—During this phase, it creates important background knowledge for patent law design, as well as issues such as financial controls placement, control structure design, controller implementation, and embedded controller structures.

HOW THE TOOL IS USED

The financial reporting models and outputs are the lens through which investors perceive and understand the wealth-generating innovation activities of a company, along with the results of those activities. The business case for innovation will succeed or fail based on its capacity to communicate these activities, especially innovation-related outcomes, clearly and completely. Businesses continually evolve, entering or leaving markets, developing new products and services, and finding new ways to attract and retain customers. These changing business practices require a basic understanding of financial reporting, and how the financial reporting model evolve as well. A basic understanding is needed by the innovator concerning basic reports and what the various columns may mean, so that the organization can always meet investors' needs for the information required to evaluate investments and make financial decisions.

A balance sheet, also known as a *statement of financial position*, reveals a company's assets, liabilities, and owners' equity (net worth). The balance sheet, together with the income statement and cash flow statement, make up the cornerstone of any company's financial statements. If you are a shareholder of a company, it is important that you understand how the balance sheet is structured, how to analyze it, and how to read it.*

---

* Read more at http://www.investopedia.com/slide-show/reading-the-balance-sheet/#ixzz3XQLc9f18.

Other Definitions

Assets—Assets are any property owned by a person or business. Tangible assets include money, land, buildings, investments, inventory, cars, trucks, boats, or other valuables. Intangibles such as goodwill are also considered to be assets.

Balance sheets—The balance sheet provides investors with information about a company's assets and the claims against those assets.

Budget—A budget is a plan that outlines an organization's financial and operational goals. Thus, it may be thought of as an action plan; planning a budget helps a business allocate resources, evaluate performance, and formulate plans.

Cash flow statements—One of the quarterly financial reports any publicly traded company is required to disclose to the Securities and Exchange Commission and the public. The document provides aggregate data regarding all cash inflows a company receives from both its ongoing operations and external investment sources, as well as all cash outflows that pay for business activities and investments during a given quarter. (*Source: Investopedia.*)

Income statement—A financial statement that measures a company's financial performance over a specific accounting period. Financial performance is assessed by giving a summary of how the business incurs its revenues and expenses through both operating and nonoperating activities. It also shows the net profit or loss incurred over a specific accounting period, typically over a fiscal quarter or year. (*Source: Investopedia.*)

STARTING POINT

The *annual operating plan* is a formal statement of business short-range goals, and reasons they are believed to be attainable, the plan for reaching these goals, and the funding approved for each part of the organization (budget). It includes the implementation plan for the coming years (years 1–3) in the strategic plan. It may also contain background information about the organization or team attempting to reach these goals. One of the end results is performance planning for each manager and employee who will be implementing the plan over the coming year. The annual operating plan is often referred to as just the operating plan (OP). One of the key elements in the annual operating plan is the budgets for the organization. From an organizational standpoint, the budget is based on two major sections. They are as follows:

1. Income projections
   - Sales revenue
   - Interest income
   - Investment income
   - Other income
2. Expense projections
   - Accounting services
   - Advertising
   - Bank service charges
   - Consulting charges
   - Credit card fees
   - Delivery charges
   - Deposits for utilities
   - Estimated taxes
   - Health insurance
   - Hiring costs
   - Installation/repair of equipment
   - Interest on debt
   - Inventory purchases
   - Legal expenses
   - Licenses/permits
   - Loan payments
   - Office supplies
   - Payroll
   - Payroll taxes
   - Printing
   - Professional fees
   - Rent/lease payments
   - Retirement contributions
   - Subcontracting expenses
   - Subscriptions and dues
   - Supplier charges
   - Utilities and telephone
   - Vehicle expenses
   - Other

The previous list shows typical line items that might be included in a budgeting system. It is not meant to be a complete list nor is it meant to be items that must be reported in all budgets. The budgeting process needs to be adjusted to the needs of the total organization with individual parts of the organization selecting the relevant line items to be included in their specific budget. The individual budgets plus the OP serve as key inputs in designing the financial reporting status for the organization.

Since the 19th century, operating planning and budgeting cycles had been conducted annually, either starting January 1 and ending December 31 of the same year, or based on the organization's fiscal start year. This is been a tried and proven approach. Even the U.S. federal government operates under these tried-and-proven approach. It is considered a best practice to prepare an annual plan and supporting budget each year and stick to it. At IBM, managers were expected to manage their annual budget within plus or minus 5%. We believe that the annual budgeting cycle is a right approach for many organizations; however, for a few leading-edge organizations, it may be time for a change.

How can you manage a business in the 21st century using 19th-century planning and budgeting approaches?

H. James Harrington

In this fast-changing business environment, we need to be able to quickly react to new opportunities as they arise. To take maximum advantage of the fast-changing environment, some organizations have abandoned the annual planning and budgeting cycle in favor of a rolling forecast—an action plan that is used to update the rolling operating plan. To accommodate the older conservative proven approach and the new more risky rolling operating plan approach, we present both approaches. Both of these approaches work, but we like the rolling operating plan approach innovative programs better. We realize that many of the successful business leaders and chief financial officers will feel much more comfortable with the tried-and-proven best practices of the annual operating plan, but we ask you to at least consider that there is another way to control your organization.

There are a number of excellent software programs available to assist in the budgeting cycle and the financial reporting cycle. For a small company, one of the commonly used is QuickBooks.

FINANCIAL REPORTING FOR INNOVATION

Financial reporting/management in *innovation* is more than a tool; it is methodology that can be used in the design and understanding of reporting systems to ensure that they perform consistently in the hands of those who need and use the information. It involves reporting more frequently than quarterly, and basing financial statements on observed current market values, instead of allocated and assumed amounts derived from original and irrelevant costs. It recognizes that financial reporting standards can only establish minimum requirements.

Innovators realize that users want more information and need to grasp that they are not constrained against exceeding the minimums. You need to voluntarily publish more informative financial reports more frequently. Besides changing how preparers approach reporting, this financial reporting big idea would influence how standard setters and other regulators carry out their responsibilities.* The four key areas where *positive innovation* can provide opportunity are (a) financing and growing the private economy; (b) promoting inclusiveness; (c) increasing efficiency, access, and the customer experience; and (d) rebalancing risk across sectors of the economy.

The things that the innovator needs to know about financial accounting are summarized as follows (see Figure 11.1):

1. *The balance sheet equation is assets = liabilities + shareholders' equity.* This equation means that assets, which are the means used to operate the company, are balanced by a company's financial obligations along with the equity investment brought into the company, plus its retained earnings. The total assets must equal the liabilities plus the equity of the company.
2. *Understand the meaning of current assets.* Current assets are most often known to have a life span of one year or less, meaning they

---

* Financial innovation has come under significant scrutiny over the past decade, and nobody can argue that certain financial innovations went badly wrong in the run up to the recent crisis of 2009. Nevertheless, *positive innovation* continues to be needed to address the challenges society will be facing in the future, outlines the report "Rethinking financial innovation, reducing negative outcomes while retaining the benefits" (World Economic Forum, written in collaboration with Oliver Wyman). This report explores the topic of innovation in the financial services industry and its effect on the wider economy. For details, see http://www3.weforum.org/docs/WEF_FS _RethinkingFinancialInnovation_Report_2012.pdf.

1. The balance sheet equation. The main formula behind balance sheets is Assets = Liabilities + Shareholders' equity.
2. Know the current assets.
3. Know the noncurrent assets.
4. Learn the different liabilities.
5. Learn about shareholders' equity.
6. Analyze with ratios.

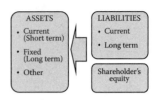

FIGURE 11.1
Financial accounting basic terms.

can be readily and easily converted into cash. Such assets include cash and cash equivalents, accounts receivable and inventory, and it is important to be familiar with these terms. Cash, the most fundamental of current assets, also includes nonrestricted bank accounts and checks. Cash equivalents are very safe assets that can be readily converted into cash (i.e., U.S. Treasuries are one example). Cash equivalents are also distinguished from other investments through their short-term existence—they mature within 3 months—whereas short-term investments are 12 months or fewer, and long-term investments are any investments that mature in excess of 12 months. Accounts receivables consist of the short-term obligations owed to the company by its clients. Companies often sell products or services to customers on credit; these obligations are held in the current assets account until they are paid off by the clients.

3. *Know the noncurrent assets.* Noncurrent assets are assets that are not turned into cash easily, and are expected to be turned into cash within a year (or have a life span of more than a year). They can refer to tangible assets such as machinery, computers, buildings, and land. Noncurrent assets also can be considered as intangible assets, such as goodwill, patents, or copyright. While these assets are not physical in nature, they are often the resources that can help to really make or break a company—the value of a brand name, for instance, should not be underestimated by innovation practitioners. Depreciation is calculated and deducted from most of these assets, which represents the economic cost of the asset over its useful life.

4. *Learn the different liabilities.* On the other side of the balance sheet are the liabilities. These are the financial obligations a company owes to outside parties. Like assets, they can be both current and long

term. Long-term liabilities are debts and other nondebt financial obligations, which are due after a period of at least one year from the date of the balance sheet. Current liabilities are the company's liabilities that will come due, or must be paid, within one year. This is includes both shorter-term borrowings, such as accounts payables, along with the current portion of longer-term borrowing, such as the latest interest payment on a ten-year loan.

5. *Learn to talk about shareholders' equity.* Shareholders' equity is the initial amount of money invested into a business. If, at the end of the fiscal year, a company decides to reinvestt its net earnings into the company (after taxes), these retained earnings will be transferred from the income statement onto the balance sheet into the shareholder's equity account. This account represents a company's total net worth. In order for the balance sheet to balance, total assets on one side have to equal total liabilities plus shareholders' equity on the other.

6. *Analyze with ratios.* Financial ratio analysis used by innovators consists of formulas to gain insight into the company, its Intellectual Properties (IP) portfolios, and its operations. For the innovators' balance sheet, using financial ratios (like the debt-to-equity ratio) can show you have a better idea of the company's financial condition along with its operational efficiency. It is important to note that some ratios will need information from more than one financial statement, such as from the balance sheet and the income statement alike.

Managing the Negative Outcomes with Financial Acumen

It is important that conventional management systems be put in place, including the *boring-stuff* financial reporting, in order to improve entrepreneurial innovation outcomes and hold innovation practitioners accountable. Basically, the process of innovation needs to be managed in such a way that nurtures individual initiative while meeting corporate growth requirements. In many ways, some might argue that this is simply innovation management, whether the organization be a tiny start-up or a huge industrial enterprise.

Innovation by definition introduces uncertainty, or it is not innovation, which gives rise to unintentional negative consequences and outcomes. Given the financial sector's relationship to the rest of the economy, it is vital that the likelihood of negative outcomes with widespread consequences is reduced. Yet, the dynamics of the sector and of innovations themselves make it impossible to reliably predict negative outcomes for individual innovations. However, enhancements to existing governance procedures, by adapting existing risk management mechanisms and other processes, can increase sensitivity to the specific contribution of innovation to uncertainty and risk.*

Corporate financial statements and their related disclosures are fundamental to sound investment decision making. The well-being of the world's financial markets, and of the millions of investors who entrust their financial present and future to those markets, depends directly on the information financial statements, disclosures provided, and how the management interprets them and how quickly they react to negative changes. Consequently, the quality of the information drives global financial markets. The quality, in turn, depends directly on the principles and standards managers apply when recognizing and measuring the economic activities and events affecting their companies' operations.

Innovation-Centered Financial Reporting (ICFR)

Our Innovation-Centered Financial Reporting (ICFR) model provides investors with a revised set of four financial statements, with all statements being of equal importance. The first three statements provide information at a higher level of aggregation. The fourth statement provides a level of disaggregation that is missing today and that is critical to investors' understanding of how the other statements, and the items in them, articulate.

Whether one focuses on extremely damaging unintended outcomes or on lesser ones, a review of other sectors also demonstrates that essentially every industry has some type of governance mechanism that attempts to channel innovation so that society as a whole can enjoy the benefits, while exposure to negative outcomes of innovation is reduced. The governance mechanisms in financial services include extensive risk management processes that have been developed over the past decades.

---

* The World Economic Forum (WEF) report takes the position that the primary responsibility for improving the management of financial innovation lies with banks and insurers. It provides a taxonomy of potential negative outcomes and recommends initiatives for companies, industry bodies, and regulators. For institutions, it recommends improvements to existing enterprise risk management techniques, new product impact assessments, better design of incentives, and enhanced *consumer orientation*. The authors believe these changes can materially reduce the odds of unintended negative consequences from innovation.

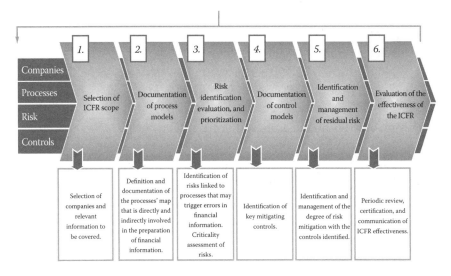

FIGURE 11.2
Innovation-centered financial reporting model.

Initially focused on credit and interest rate risk, in banking, and on actuarial risk in insurance, the risk management frameworks in financial services have gradually extended their scope to address myriad additional categories of risk, including reputational risk, event risk, operational risk, and others.* Aside from explicit risk management frameworks, governance mechanisms also include new product development and approval processes employing various safeguards against unwise innovation; and of course, they include an extensive regulatory infrastructure that has been in place since before the crisis and is already being enhanced, as illustrated in Figure 11.2.

The ICFR model was developed by Harrington Management Systems and BBVA Group's management, in accordance with international standards established by the Committee of Sponsoring Organizations of the Treadway Commission (known as COSO). This model stipulates five components that must form the basis of innovation effectiveness and efficiency of systems of internal control and financial reporting:

1. Establishment of an appropriate control framework to monitor these activities
2. Assessment of all of the risks that could arise during the preparation of financial information

---

* See footnote on page 152.

3. Designing the necessary controls to mitigate the most critical risks
4. Establishment of an appropriate system of information flows to detect and report system weaknesses or flaws
5. Monitoring of the controls to ensure they perform correctly and are effective over time

This set of ICFR financial reporting statements includes

1. *Balance sheets*—Minimum of one year, and in some cases three years (three balance sheets and two income and cash flow statements provide a full two years of data), with accounts listed in order of decreasing liquidity within each category.
2. *Cash flow statements*—Minimum of one or two years, prepared using the direct method, with a supplemental schedule of significant noncash financing and investing activities.
3. *Statements of changes in net assets available to common shareowners*—Minimum of two years, that
   • Identify and distinguish among current-period cash and accrual transactions, estimates, and changes in the fair values of balance sheet accounts
   • Provide information by nature of each resource consumed rather than the function for which it is consumed
   • Display transactions with owners that affect net assets, such as dividends and new share issuances
4. *Reconciliation of financial position*—Reconciles the comparative balance shets by further disaggregating the amounts in the statements mentioned previously and by clearly showing how the statements articulate.

Description of the Balance Sheet

The balance sheet provides investors with information about a company's assets and the claims against those assets

• Resources available to it
• Relative liquidity of the resources
• Recognized and contingent claims against those resources
• Relative time to maturity of these claims

The importance of the balance sheet to investment decision making has diminished in recent decades. Arguably, this is a result of such factors as

- Models that focus on reported earnings and cash flows rather than accounting net worth
- The reduced relevance of historical cost-based numbers for decision making given both general inflation and price changes for specific assets and liabilities
- The omission from the balance sheet of major classes of resources and obligations (e.g., intangible assets and operating leases and other off-balance-sheet financing arrangements)
- Accounting choices that limit comparability

It is time to refocus attention on the balance sheet, to correct its deficiencies, and to restore its usefulness as a communicator of essential information about companies and their innovation impacts and efforts (see Table 11.1).

The following definitions will help the innovator understand the balance sheet for the activities he or she is responsible for.

- *Accounts payable*: An accounting entry that represents an entity's obligation to pay off a short-term debt to its creditors.
- *Accounts receivable*: Money that is owed to a company by a customer for products and services provided on credit. This is often treated as a current asset on a balance sheet. A specific sale is generally only treated as an account receivable after the customer is sent an invoice.
- *Accrued liabilities*: Accrued liabilities are liabilities that reflect expenses on the income statement that have not been paid or logged under accounts payable during an accounting period; in other words, obligations for goods and services provided to a company for which invoices have not yet been received.
- *Accrued pension liability*: An accounting term for an expense to pay its employees at some point in the future for benefits earned under a pension plan that has not yet paid.
- *Accumulated depreciation*: A long-term *contra asset account* (an asset account with a credit balance) that is reported on the balance sheet under the heading Property, Plant, and Equipment.
- *Advances from customer*: A liability account used to record an amount received from a customer before a service has been provided

TABLE 11.1

Example of a Typical Balance Sheet

	31December	
	$20X_3$	$20X_4$
Assets		
Cash	€4,000,000	€5,918,411
Marketable securities	0	196,100
Accounts receivable	595,000	845,000
Less: Allowance for had debts	(20,000)[a]	(70,500)
Net Accounts Receivable	**575,000**	**774,500**
Inventory	850,000	619.694
Leased asset	0	25,756
Investment in affiliate	0	722.250
Building	3,600,000	4,260,000
Less: Accumulated depreciation	(100,000)	(275.000)
Net Building	**€3,500,000**	**€3,985,000**
Total Assets	**€8,925,000**	**€12,241,711**
Claims and Common Shareowners' Equity		
Accounts payable	€850,000	€375,000
Accrued liabilities	28,000	78,000
Advances from customers	15,000	190,000
Interest payable	0	125,000
Income taxes payable	54,630	75,451
Dividends payable	0	35,000
Current portion of lease liability	0	9,208
Short-term debt	0	500,000
Accrued stock compensation	6,000	13,500
Lease liability	0	24,870
Deferred income liability	S3,60S	56,819
Accrued pension liability	2,400	4,800
Bonds payable	0	2,500,000
Minority interest	100,000	100,000
Perpetual preferred stock	300,000	300,000
Total Claims	**€1,379,738**	**€4,387,648**
Common stock	€600,000	€600,000
Additional paid-in capital	4,000,000	4,000,000
Treasury stock	(100,000)	(100,000)
Retained net assets	3,045,262	3,354,063
Total Common Shareowners' Equity	**€7,545,262**	**€7,854,063**
Total Claims and Common Shareowners' Equity	**€8,925,000**	**€12,241,711**

[a] Parentheses indicate negative numbers.

or before goods have been shipped. This account is referred to as a deferred revenue account and could be entitled Customer Deposits or Unearned Revenues.

- *Additional paid-in capital*: A value that is often included in the contributed surplus account in the shareholders' equity section of a company's balance sheet. The account represents the excess paid by an investor over the par-value price of a stock issue. Additional paid-in-capital can arise from issuing either preferred or common stock.
- *Allowance for bad debt*: A valuation account used to estimate the portion of a bank's loan portfolio that will ultimately be uncollectible. When a loan goes bad, the asset is removed from the books and the allowance for bad debt is charged for the book value of the loan.
- *Asset*: In financial accounting, an asset is an economic resource. Anything tangible or intangible that is capable of being owned or controlled to produce value and that is held to have positive economic value is considered an asset.
- *Bonds payable*: Generally a long-term liability account containing the face amount, par amount, or maturity amount of the bonds issued by a company that are outstanding as of the balance sheet date.
- *Building*: A noncurrent or long-term asset account that shows the cost of a building (excluding the cost of the land). Buildings will be depreciated over their useful lives by debiting the income statement account Depreciation Expense and crediting the balance sheet account Accumulated Depreciation.
- *Cash*:
  - Money in the form of coins or banknotes, especially that issued by a government.
  - Money or an equivalent, as a check, paid at the time of making a purchase.
- *Claims*:
  - Legal demand or assertion by a claimant for compensation, payment, or reimbursement for a loss under a contract, or an injury due to negligence.
  - Amount claimed by a claimant.
- *Common shareholders' equity*: A firm's total assets minus its total liabilities. Equivalently, it is share capital plus retained earnings minus treasury shares. Shareholders' equity represents the amount by which a company is financed through common and preferred

shares. If a corporation has issued only one type, or class, of stock, it will be common stock.

- *Common stock*: The type of stock that is present in every corporation. (Some corporations have preferred stock in addition to their common stock.) Shares of common stock provide evidence of ownership in a corporation. Holders of common stock elect the corporation's directors and share in the distribution of profits of the company via dividends. If the corporation were to liquidate, the secured lenders would be paid first, followed by unsecured lenders, preferred stockholders (if any), and lastly the common stockholders.
- *Current portion of lease liability*: This is the portion of a long-term capital lease that is due within the next year.
- *Deferred income tax liability*: Generally Accepted Accounting Principles allow management to use different accounting principles or methods for reporting purposes than it uses for corporate tax filings (Internal Revenue Service [IRS]). Deferred tax liabilities are taxes due in the future (future cash outflow for taxes payable) on income that has already been recognized for the books. In effect, although the company has already recognized the income on its books, the IRS lets it pay the taxes later (due to the timing difference). If a company's tax expense is greater than its tax payable, then the company has created a future tax liability (the inverse would be accounted for as a deferred tax asset).
- *Dividends payable*: A dividend is a distribution of a portion of a company's earnings, decided by the board of directors, to a class of its shareholders.
- *Income taxes payable*: A type of account in the current liabilities section of a company's balance sheet. This account comprises taxes that must be paid to the government within 1 year.
- *Interest payable*: A current liability account that is used to report the amount of interest that has been incurred but has not yet been paid as of the date of the balance sheet.
- *Inventory*:
  - A company's merchandise, raw materials, and finished and unfinished products that have not yet been sold. These are considered liquid assets, since they can be converted into cash quite easily. There are various means of valuing these assets; however, to be conservative, the lowest value is usually used in financial statements.

- The securities bought by a broker or dealer in order to resell them. For the period that the broker or dealer holds the securities in inventory, he or she is bearing the risk related to the securities, which may change in price.
- *Investment in affiliate asset*: Investment in affiliate (or equity investment) is the noncurrent asset account on the balance sheet that serves as a proxy for company A's economic interest in company B's assets and liabilities.
- *Lease liability*: This is a written agreement under which a property owner allows a tenant to use and rent the property for a specified period. Long-term capital lease obligations are net of current portion.
- *Leased asset*: A finance lease or capital lease is a type of lease. It is a commercial arrangement where the lessee (customer or borrower) will select an asset (equipment, vehicle, software), the lessor (finance company) will purchase that asset, and the lessee will have use of that asset during the lease.
- *Market securities*:
  - Very liquid securities that can be converted into cash quickly at a reasonable price.
  - Marketable securities are very liquid as they tend to have maturities of less than one year. Furthermore, the rate at which these securities can be bought or sold has little effect on their prices.
- *Minority interest*: In accounting, minority interest (or noncontrolling interest) is the portion of a subsidiary corporation's stock that is not owned by the parent corporation. The magnitude of the minority interest in the subsidiary company is generally less than 50% of outstanding shares; otherwise, the corporation would generally cease to be a subsidiary of the parent.
- *Perpetual preferred stock*: A type of preferred stock that has no maturity date. The issuers of perpetual preferred stock will always have redemption privileges on such shares. Issued perpetual preferred stock will continue paying dividends indefinitely.
- *Retained net assets*:
  - The percentage of net earnings not paid out as dividends but retained by the company to be reinvested in its core business, or to pay debt. It is recorded under shareholders' equity on the balance sheet.

- The formula calculates retained earnings by adding net income to (or subtracting any net losses from) beginning retained earnings and subtracting any dividends paid to shareholders.
- *Short-term debt*: An account shown in the current liabilities portion of a company's balance sheet. This account comprises any debt incurred by a company that is due within one year. The debt in this account is usually made up of short-term bank loans taken out by a company.
- *Treasury stock*: The portion of shares that a company keeps in their own treasury. Treasury stock may have come from a repurchase or buyback from shareholders, or it may have never been issued to the public in the first place. These shares do not pay dividends, have no voting rights, and should not be included in shares outstanding calculations.

Description of Cash Flow Statements

A cash flow statement, when used in conjunction with the rest of the financial statements, provides information that enables users to evaluate the changes in net assets of an enterprise, its financial structure (including liquidity and solvency), and its ability to affect the amounts and timing of cash flows in order to adapt to changing circumstances and opportunities. Cash flow information is useful in assessing the ability of the enterprise to generate cash and cash equivalents, and enables users to develop models to assess and compare the present value of the future cash flows of different enterprises. It also enhances the comparability of the reporting of operating performance by different enterprises because it eliminates the effects of using different accounting treatments for the same transactions and events.*

From an innovation financial management perspective, we made clear that we believe that

- All assets available to the company and all obligations and commitments that can consume the net assets available to investors should be recorded in the balance sheet.
- All balance sheet assets and liabilities should be reported at fair value, starting with all financial instruments as soon as possible.
- All balance sheet items should be disaggregated.

---

* The International Accounting Standards Board, in International Accounting Standard (IAS) 7, describes some of the benefits of cash flow information: IAS 7, *Cash Flow Statements*, effective January 1, 1994, paragraph 4.

- Assets and liabilities should not be netted.
- Income statement components and related balance sheet items should be reported by nature rather than function.
- Related assets and liabilities should be classified together by category.
- All assets and liabilities should be ordered by liquidity, specifically by decreasing liquidity, within their respective categories.

On Using QuickBooks

For those innovators who are involved in an already established organization, the financial management structure probably is already in place. All the innovator needs to do is to understand the terminology, rules, and practices presently employed throughout the organization. Unfortunately, many of the individuals involved in innovation are often working out of their garage. The start-up innovators need to establish a completely new comprehensive financial management system. There are many programs available that will aid in establishing an initial financial management system. We will use one of these, called QuickBooks, to demonstrate what is required to establish a new simplified financial management system into a small start-up organization.

It is not necessary to be an accountant to enjoy and utilize its capabilities. It is necessary to have an understanding of basic accounting principles, requirement procedures, and jargon to avoid frustration that can arise from incorrect posting or data entry. It is also necessary to equate procedures used in QuickBooks with standard data entry and accounting processes and steps.

Accounting programs are of two types. The first of these, single entry, operate like checkbook- or ledger-type systems, since they require only a single entry in a specific ledger or account. They require less skill to use, but may not accurately portray the true financial position of the firm at a point in time. The second type, double entry, follows commonly accepted accounting rules.

For every debt, there must be an offsetting credit and vice versa. Firms with complex inventory management, payroll, and an extensive number of enterprises are well advised to choose a double-entry system, either because of the dictates of the law or practical necessity.

QuickBooks is a double-entry system but most of the heavy lifting is accomplished by the program since multiple offsetting entries are accomplished by the program. Having said that, it is important to "get it right the

TABLE 11.2

QuickBooks Financial System

QuickBooks System Modules	
• Chart of accounts	• Reports
• Cash receipts and expenditures	✓ Balance sheet
• Capital and depreciation schedules	✓ Profit and loss
• Asset and liability schedules	✓ Purchases
• Receivable and payable schedules with aging	✓ Sales
• Inventory schedules	✓ Inventory
• Financial statements	✓ Budgets
• Payroll	✓ Custom

first time" when entering data, because backing out incorrect entries can, in some cases, be tedious and time consuming* (see Table 11.2).

The flow chart in Figure 11.3 portrays the composition of a typical computerized accounting program. In general, data entry is accomplished through the utilization of a variety of modules, as shown in Figure 11.3. Figure 11.3 reflects a good financial accounting program that comprises connected modules containing information an innovation manager needs for financial reporting, diagnosis, control financial planning, and financial management.

For Normal Financial Reporting

1. The ultimate objective of the wealth-generation process is to generate cash. Thus, it is critically important for innovators to understand how companies generate cash and how they manage cash receipts and payments. Innovators will achieve this understanding only when the business reporting model communicates the process clearly and completely. Innovators need a reporting model that better communicates the risks and uncertainties implicit in the events and transactions, and thus in the amounts recognized in the financial statements.

2. This information should directly support the development of forward-looking estimates of sustainable cash flows. Academic research confirms that disaggregating earnings into cash flow and the

---

* From a brief called QuickBooks Notes, courtesy of Fresno State University. For details, see http://zimmer.csufresno.edu/~jamesco/QbooksIntro.pdf.

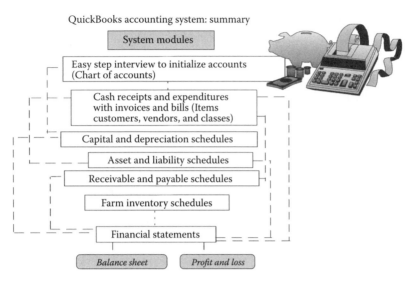

QuickBooks accounting system: summary

System modules

Easy step interview to initialize accounts
(Chart of accounts)

Cash receipts and expenditures
with invoices and bills (Items
customers, vendors, and classes)

Capital and depreciation schedules

Asset and liability schedules

Receivable and payable schedules

Farm inventory schedules

Financial statements

Balance sheet   *Profit and loss*

FIGURE 11.3
QuickBooks financial accounting reporting system.

components of accruals enhances earnings' predictive ability relative to aggregate earnings. It is not surprising that disaggregating
the earnings series would improve predictive ability, a primary purpose of financial reporting, because information lost in the aggregation process is preserved in the separate series. In addition, separate
reporting of cash flows and accruals permits innovators with different analysis and investment objectives to apply different weights to
the series in developing forecasts and valuations.

3. This research supports our call in this chapter for a new business
reporting model, one that disaggregates the jumbled income statement and cash flow numbers, and that clearly and completely communicates both cash flow information and accruals. Financial
reporting for innovation is a robust yet flexible framework to analyze the introduction, growth, and maturation of innovations, and to
understand and report financial information about the technological
cycles in a format needed and understood by the innovator using the
data. The model also has plenty of empirical evidence; it was exhaustively studied within many industries, including semiconductors,
telecommunications, hard drives, photocopiers, jet engines, and so
on. Figure 11.4 illustrates the process of what it takes to climb and
jump the S-curve in business innovation, and to financially report
on the building of high performance.

Building high performance

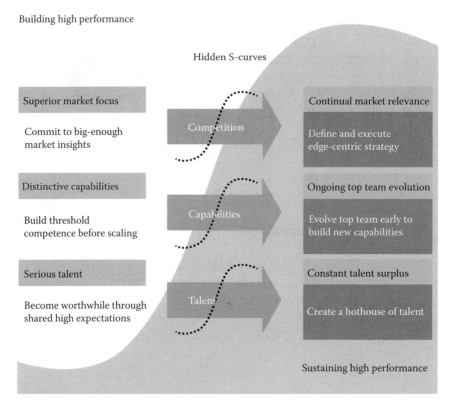

FIGURE 11.4
Financial reporting model for building innovation.

Figure 11.4 demonstrates the process for the financial reporting param-eters involved in climbing and jumping the S-curve: superior market ser-vice, distinctive capabilities, and serious talent. The balance sheets should include the outcomes and metrics for continual market relevance, ongoing operation team evaluations, and creating a talent surplus.

Definition: The S-curve is a mathematical model also known as the logis-tic curve, which describes the growth of one variable in terms of another variable over time. S-curves are found in many fields of innovation, from biology and physics to business and technology.

Recognizing the way in which innovation supports economic growth, many governments worldwide encourage financial reporting of the investment in research and development by allowing companies to claim tax credits for the amount spent on it, particularly for technology-driven innovations.* The power of innovation derives from its combination with investment and competition. Innovation initially benefits the innovator and investment magnifies the returns. Competition then helps distribute the benefits of innovation more widely across society, driving down prices and making new products and services widely available. Some innovations prove to be what are called general-purpose technologies, upon which a myriad of further innovations can be built. Electricity generation is a 19th-century example of such an innovative wellspring, transistors and microchips are 20th-century examples, and the Internet is a modern one.

For the past 5 years, a World Economic Forum team has worked with many constituents and with the active support of Oliver Wyman to analyze the relevant literature and seek the counsel of more than 100 businesses, and political and academic leaders around the globe.† The project's findings converge on a core theme, namely that the most important aspect of innovation, in the context of risk, is that there are no historical metrics to determine its impact on the world. On the other hand, innovation done right means revenue gains, which can be measured and acted upon as a result of good financial reporting tools. In addition to the broadly visible *front end* of platforms designed for innovators, financial reporting innovation management platforms also provide a backroom or *innovation dashboard*, where leaders can track and analyze the innovations occurring in the organizational work units. Figure 11.5 graphically shows how innovation within an organization drives increased revenue.

A typical dashboard shows leaders how companies that perform well in innovation initiatives grow three times faster than lesser-performing companies. It also shows that many ideas or value propositions are created over time, which departments or groups contribute, which groups comment and make innovations stronger, how many ideas turn into pilot projects, and how many become commercially viable products or services in the marketplace.

---

* Hall, B. Tax incentives for innovation in the United States. ZABALA-Spain: Asesoria Industria, Report to the European Union. Available at: http://elsa.berkeley.edu/~bhhall/papers/BHH01%20 EU%20Report%20USA%20rtax.pdf, 2001.
† The project was conducted under the stewardship of a steering committee, chaired by Stefan Lippe, the former group chief executive officer of Swiss Re, supported by a working group of senior industry executives, regulators, and academics who guided its work.

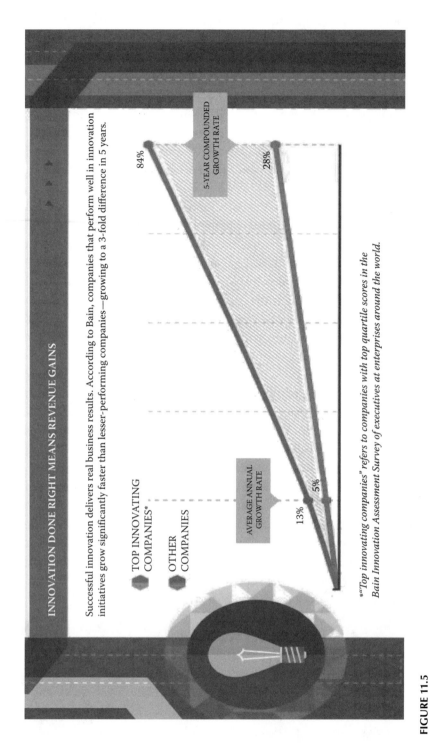

FIGURE 11.5

Innovation done right means revenue gains.

Therefore, the industry needs to pay special attention to the ways in which its mechanisms for assessing and managing risk should be adapted to take better financial reporting account of innovations. Another important finding is that the postlaunch management of an innovation, including *downstream variants* and new applications, is more relevant, in many cases, than the original innovation itself. Financial reporting looks to ensure that our future is one of inspired collaboration and bold solutions to the global, regional, and industry challenges, and not a return to the status quo.

Comments Related to Innovation

1. A co-founder of PayPal and one of the most important futurists in the United States today, Thiel assesses whether interviewees have the potential to work passionately at innovating by asking them, "What important truth do very few people agree with you on?" When he says *very few people*, he means a really small number because he believes real innovation is unlikely without a unique perspective on truth different from the status quo. His own view is, well, innovative, because it involves two kinds of progress. He calls one of them *horizontal*, by which he means new things only improve on previously existing old things, such as electric typewriters replacing manuals, and spinning type balls superseding type bars. He characterizes this progress as *one to n*.

2. Another example is globalization, which Thiel says has merely tried to replicate what worked in the United States in the 19th and 20th centuries. He observes that it is just not possible to recreate American lifestyles in countries with less real estate and more people crammed into small spaces. In contrast, he uses the terms *vertical* and *zero to one* to describe those innovations that have the greatest potential for progress because they are more likely to create revolution, instead of evolution. One example is word processing, because it did not merely improve typewriters but virtually eliminated them, while significantly changing the way ideas are expressed and conveyed in text.

3. Finally, in the move from passenger trains to airplanes, for decades the railroads kept on innovating and improving their services, but seemingly failed to comprehend how much faster airlines could move their customers. Although some still travel long distance

by rail, their numbers are small and the fares must be subsidized. Notably, innovation and progress are never automatic. Those who strive to achieve them must be courageous and willing to do tough work in the face of opposition and ridicule.

Financial Reporting Capability

Financial management capability refers to the combination of knowledge, understanding, skills, attitudes, and especially behaviors that innovators (and people) need to make sound personal finance decisions suited to their organizational, social, and financial circumstances. Some of the most important behaviors for financial capability include

1. Making ends meet
2. Keeping track of one's finances
3. Planning ahead
4. Choosing financial products wisely
5. Staying informed about financial matters, often also termed *getting help*

In an environment of rapid financial innovation, these behaviors take on even more importance.

So what do these ideas mean for financial accountants? We think this viewpoint supports our long-held position that financial reporting has been largely stagnant over the last hundred years. Basically, we fear accountants have plodded along horizontally without significantly improving financial reporting's impact on capital markets, economies, and eventually society itself. Sure, they have embraced new data processing technologies, but they have applied them only to doing the same things they have always done.

"The fundamental impulse that sets and keeps the capitalist engine in motion comes from the new consumer goods, the new methods of production or transportation, the new markets, the new forms of industrial organization that capitalist enterprise creates .... This process of Creative Destruction is the essential fact about capitalism." Innovation always changes the status quo, but some innovations cause greater disruption than others. In the most severe cases, radical innovations fundamentally change society and spawn further generations of innovation. At the other end of the spectrum, incremental innovations help to differentiate a company from its competitors and, for the consumer, offer a constant round of useful improvements to existing products, processes, and services, as well as to reductions in real prices.

Advantages

- Skilled innovation practitioners who understand financial accounting and how to use them have a powerful control tool with which to attain organizational innovation goals.
- The major strength of financial management is that it coordinates activities across departments.
- Budgets translate strategic plans into action. They specify the *resources*, *revenues*, and *activities* required to carry out the strategic plan for the coming year.
- Financial accounting provides an excellent record of organizational activities and improves communication with employees.
- Financial management results in improve resources allocation, because all requests are clarified and justified.
- Financial accounting for innovation provides a tool for corrective action through reallocations.

Disadvantages

- The major problem occurs when budgets are applied mechanically and rigidly.
- Budgets can demotivate employees because of lack of participation. If the budgets are arbitrarily imposed top down, employees will not understand the reason for budgeted expenditures, and will not be committed to them.
- Budgets can cause perceptions of unfairness.
- Budgets can create competition for resources and politics.
- A rigid budget structure reduces initiative and innovation at lower levels, making it impossible to obtain money for new ideas.
- These dysfunctional aspects of financial management systems may interfere with the attainment of the organization's innovation goals.

SUMMARY

What drives us to make things better? Why do we push harder to improve a process, a customer journey, to make continuous leaks to our financial reporting ... what motivates us to think differently and drive innovation? Whether you are someone on the financial front lines or the founder of a tech start-up, we are dealing with these, your customers, your stakeholders, your team members, your clients—the ones that make or break your new product or service innovation, your company, your profit, and your underlying ability to pursue all of your creative dreams and passions. Innovators that do not understand and manage the financial system related to their new and creative ideas are the ones that have caused from 40% to 60% of the creative projects to fail.

That is what we are doing by innovating in the financial space—helping people pursue their life's passions by removing friction and ambiguity from our financial reporting. Adopting financial management systems for innovation will give you comparability, increased audit efficiency, reduced information misunderstandings, and price savings as more economic activities become globalized. On the other hand, it could eliminate, in some cases, incentives to innovate. In other words, the standard is affected since compromises need to be designed to achieve consensus because of various political pressures and economic interest.

EXAMPLES

Figure 11.6 is an example of a balance sheet.
Figure 11.7 is an example of a profit and loss statement.

SOFTWARE

Some of the commercial software available include but are not limited to

- QuickBooks: http://quickbooks.intuit.com/.

Balance Sheet

03/08/11

Cash Basis

As of December 31, 2010

	Dec 31, 10
ASSETS	
Current Assets	
Checking/Savings	
Checking	278.44
Total Checking/Savings	278.44
Total Current Assets	278.44
TOTAL ASSETS	**278.44**
LIABILITIES & EQUITY	
Liabilities	
Long Term Liabilities	
███████	114,500.00
Total Long Term Liabilities	114,500.00
Total Liabilities	114,500.00
Equity	
Retained Earnings	-44,817.55
Net Income	-69,404.01
Total Equity	-114,221.56
TOTAL LIABILITIES & EQUITY	**278.44**

FIGURE 11.6
Example of a balance sheet.

- Netsuite: http://www.netsuite.com/portal/seo-landing-page/accounting-4/main.shtml?gclid=CjwKEAjwr6ipBRCM7oqrj6O30jUSJACff2WHot8t3Qfe8JnzaH-YPWXDxOJe3_sT9u4isvr3fPAqRoC0a_w_wcB.
- Infor's CloudSuite Financials product line—a horizontal group of written-from-scratch applications—will likely trigger a wave of new innovation and product release activity over the next year or so. See http://www.zdnet.com/article/why-infors-cloudsuite-heralds-innovation-wave-in-the-financial-software-market/.

Profit & Loss
January through December 2010

	Jan - Dec 10
Ordinary Income/Expense	
Income	
Consulting Income	7,475.00
Total Income	7,475.00
Expense	
Automobile Expense	399.99
Bank Service Charges	243.56
Dues and Subscriptions	149.00
Interest Expense	
Finance Charge	252.08
Total Interest Expense	252.08
Licenses and Permits	340.00
Miscellaneous	623.96
Office Supplies	1,115.52
Outside Services	5,150.00
Payroll Expenses	62,379.00
Postage and Delivery	485.65
Professional Fees	
Accounting	1,840.39
Total Professional Fees	1,840.39
Taxes	
State	800.00
Total Taxes	800.00
Telephone	2,484.86
Travel & Ent	
Travel	615.00
Total Travel & Ent	615.00
Total Expense	76,879.01
Net Ordinary Income	-69,404.01
Net Income	**-69,404.01**

FIGURE 11.7
Example of a profit and loss statement.

- Software Advice: http://www.softwareadvice.com/accounting /business-accounting-software-comparison/top-ten/?utm_source =google-search&utm_medium=ppc&utm_term=%2Bfinancial %20%2Baccounting%20%2Bsoftware&utm_matchtype=b&network =g&device=c&adpos=1t3&utm_campaign=accounting_general&ad =41729697457.

12

Focus Group

H. James Harrington

CONTENTS

DEFINITION

A focus group is a structured group interview of typically 7 to 10 individuals who are brought together to discuss their views related to a specific business issue. The group is brought together so that the organizer can gain information and insight into a specific subject or the reaction to a proposed product. The information gained from focus groups aids the organization conducting the interview to make better educated decisions regarding the topic being discussed.

USER

The focus group is made up of a group of individuals that are knowledgeable of or would make use of the subject being discussed. The facilitator is used to lead the discussions and record key information related to the discussions.

OFTEN USED IN THE FOLLOWING PHASES OF THE INNOVATIVE PROCESS

The following are the seven phases of the innovative cycle. An X after the phase name indicates that the tool/methodology is used during that specific phase.

- Creation phase X
- Value proposition phase X
- Resourcing phase
- Documentation phase
- Production phase
- Sales/delivery phase X
- Performance analysis phase X

TOOL ACTIVITY BY PHASE

- Creative phase—During this phase, focus groups are often used to help identify unfulfilled needs and to evaluate how different alternatives will be viewed by external customers.
- Value proposition phase—During this phase, focus groups are often used to define the value added to the individuals that will be receiving the output from the innovative concept.
- Sales/delivery phase—Focus groups provide an excellent way to predetermine how the output from the innovative concept should be advertised and delivered to the external consumer.

- Performance analysis phase—Focus groups can be used to determine how the customer perceives the output from the innovative process. For example, will they recommend the product to their friends, and will they come back and deal with the organization again based on their experience?

HOW TO USE THE TOOL

The purpose of focus groups is to find information about how specific demographics feel about a particular topic. This information can help improve or predict future acceptance of proposed products and services. Planning and conducting a focus group is a very challenging activity. It requires a great deal of careful planning and forethought in order to get the desired results. Without the proper planning and proper facilitation skills, the results that the organization gets as a result of the focus group can be very misleading. When putting together a focus group, determine the purpose of the interview, who to interview, and develop a plan and the resources needed. Understanding what question will ultimately be answered, who is most fit to answer the questions, how individuals will be compensated, and where and when the focus group will meet are prerequisites. Key agreements should be taken during the session and recorded on a flipchart or white board if possible. This ensures that there is no misunderstanding between the facilitator and the focus group members. A set of personal and private notes should be recorded that include body language, mood of participants, consistency of answers, and specificity of responses.

There are three key elements in running a successful focus group:

- Selecting individuals to make up the focus group that are knowledgeable and are interested in the subjects that will be discussed. The individual should be open-minded and not come to the focus group with preset beliefs. It is important that they are representative of the makeup of the population that you are interested in gaining information about, and how they will react to the subject/product being discussed. Realize that there is a big difference between saying that you will buy something and putting out the money to purchase the item. The group, as made up, is a good representation of the gender and ethnic background of the population that is being evaluated.

- Selecting the facilitator: The facilitator has a key role in directing the way the group functions and the results that will be obtained from the group's meetings. They not only have to listen intensely to the way individuals in the group responded to questions but also to their nonverbal communication patterns. They also are required to have an outgoing personality that builds confidence, in the process influencing the members of the group to be at ease. They need to be very effective at asking open-ended questions, bringing out additional information related to how the group feels about the subject being discussed, and bringing out feelings that are representative of the general population that they represent. Choosing a facilitator (moderator) involves finding someone skilled in the group processes.

- The meeting agenda and questions: Usually the focus group will meet for a short period of time so the agenda has to quickly cover the purpose of the meeting and convince the individual focus group members that their thoughts and opinions will play a very important role in directing the future activities of the organization related to the subject being discussed. The questions must be open-ended, have a logical flow and progression, and time must be allotted for unanticipated questions.

Advantages

The facilitator has a set of predefined questions that allow him or her to have direct interaction with the group and allows for additional clarification and probing into key discussion points. The group setting allows for interaction between group members and for reinforcing key points between the group numbers. It allows a group member to feel more comfortable in further discussion of negative aspects of the subject than they would discuss if they had to document it in writing. It also allows for collecting information from individuals that do not have the capability of communicating in written format (e.g., children and adults whose native language is not English). They are less expensive than conducting individual interviews.

Disadvantages

Focus groups offer less control over topics of discussion. The moderator may ask leading questions and create bias. As a result, focus groups require a skilled interviewer to create questions and to lead the discussion. Dominant individuals within the group may overshadow the whole group's opinion. The data is often harder to analyze because of chaotic information gathering.

EXAMPLE

One of our clients developed a new and slightly different type of French onion soup that was packaged in a less expensive disposable container. They decided to conduct a number of focus groups to determine

1. What positive or negative impact this will have on the present French onion soup product that they had been selling for a number of years
2. What positive or negative impact this will have on capturing a larger percentage of the French onion soup market
3. What the consumer's reaction to the new packaging would be
4. If they will need to reduce the price on the new package in order to get initial acceptance

Focus groups were scheduled for Boston, Chicago, San Jose, Atlanta, Orlando, and Dallas. Three focus groups were scheduled for each site in order to get the right demographic mix. Focus group activities were scheduled for 1 hour followed by an hour for drinks and hors d'oeuvres. The focus group team members were made up of a consultant that served as the facilitator and four of the organization's staff. The focus groups were scheduled for a room that had a two-way mirror on one wall. The four staff members stayed in the room next to where the focus group meeting to observe the focus group as it was being conducted. Following the formal part of the focus group activities, the four staff members joined the focus group for the cocktails and hors d'oeuvres. This allowed them to have personal discussions with the focus group individuals to clarify any questions the staff members may have.

The results of the series of focus group meetings indicated that there would be a negative impact on the present product (between 8% and 15%). It also indicated that the company's market share of the French onion soup market would grow from 8% to between 10% and 12%, which by far offset the negative impact that the new product would have on the organization's present French onion soup sales. The focus group results also indicated that there would be no reason to price the new product lower than the present. The proposed new packaging was perceived as being a positive impact the customer's perception of the product.

SOFTWARE

Some commercial software available includes but is not limited to

- FocusGroupit: http://focusgroupit.com

SUGGESTED ADDITIONAL READING

Krueger, R.A. *Analyzing and Reporting Focus Group Results*. Thousand Oaks, CA: Sage, 1998.
Liamputtong, P. *Focus Group Methodology: Principle and Practice*. Thousand Oaks, CA: Sage Publications, 2011.

13

Identifying and Engaging Stakeholders*

Simon Speller

CONTENTS

DEFINITION

A *stakeholder* of an organization or enterprise is someone who potentially or really influences that organization, and who is likely to be affected by that organization's activities and processes, or, even more significantly, *perceives* that they will be affected (usually negatively).

USER

The user of stakeholder mapping can be anyone in the organization who makes decisions on products, processes, and services that may have an

---

* This chapter has been produced in association with Paul Goodstadt ACIB, FRSA, FTQMC, director of Development at the Total Quality Management College, Manchester, United Kingdom.

impact outside their immediate work area or team. For more senior corporate players, the potential value to the business or organization of doing stakeholder mapping is incalculable. This may be used as a personal and professional learning tool or as part of team activities in business development, marketing, strategy formulation, and operational strategy (see Table 13.1).

OFTEN USED IN THE FOLLOWING PHASES OF THE INNOVATIVE PROCESS

The following are the seven phases of the innovative cycle. An X after the phase name indicates that this tool/methodology is used in that phase.

- Creation phase X
- Value proposition phase
- Resourcing phase
- Documentation phase
- Production phase
- Sales/delivery phase X
- Performance analysis phase

TOOL ACTIVITY BY PHASE

Although focusing on and considering the stakeholders is important in all seven phases of the innovative process, the primary focus and involvement takes place during the creative phase, with the product or service being designed and customers being invited to participate in the design activities. It also plays a very important role in evaluating the impact a proposed change will have when preparing the value proposition and acquiring resources to support the change initiative. During the sales and delivery phase, a great deal of focus is placed on identifying external stakeholders and modifying the sales and marketing strategies to be in line with the external stakeholders' specific needs.

TABLE 13.1

Mapping Matrix: User Type and Level × Mapping Tool

User Level/Type and Mapping Used	Mapping Tool Used and Possible Applications/Purposes					
	Competition	Community Impact	Innovation Evaluation	Stakeholder Communications	Change Programs	Political Support
Type of mapping used	Interest (A)	Interest (A)	Interest (A), tracking (B)	Tracking (B)	Interest (A), tracking (B)	Interest (A)
Purpose of mapping	Market research	EIA, SIA	Market testing	Community profile/status	For/against change	For/against champion
External consultancy	Yes	Yes				
Board and chair of board	Yes	Yes				Maybe a personal tool, not to share?
Chief executive officer and executive	Yes	Yes	Yes	Yes	Yes	Maybe a personal tool, not to share?
Innovator	Yes	Yes	Yes, in business plan	Maybe	Maybe	Maybe a personal tool, not to share?
Implementer/manager				Yes, as part of process/project management	Yes, a core responsibility	
Operatives/contact staff		Yes, in engagement work		Yes, in engagement work and feedback	Yes, in engagement work and feedback	

Note: (A) Power–interest quadrant matrix outlined in text; (B) communications tracking stakeholder mapping tool outlined in text; EIA, environmental impact assessment; SIA, social impact assessment.

HOW TO USE THE TOOL

The list of people and groups listed as stakeholders by organizations and enterprises varies considerably, and may include any or all of the following (e.g., as listed in Burlton, 2001; Foley et al., 1997):

- Customers, consumers, and service users
- Business owners and shareholders
- Staff
- Suppliers and supply chain partners
- Communities located near the organization's activities in production, supply, and retail
- The enterprise itself

At its most holistic level of usage, the term stakeholder has been used to highlight an organization's goal to be an environmentally and socially responsible organization, in which local communities and the wider society are viewed as (or at least described as) directly interested groups—both interested in, and affected by, the operations and use of resources of an organization.

It is therefore no surprise then, that, like many words and concepts used in management speak (e.g., *innovation*, *quality*, and *excellence*), the word *stakeholder* is considered to be loaded and potentially dangerous— it comes with a health warning attached. The more so because the everyday usage of the word *stakeholder* can be, at best, vague, and, at worst, directly misleading or manipulative as corporate communications window-dressing to internal or external audiences about not very much at all, for example, on corporate social responsibility and ethics and values statements (maybe filed under Mission or Vision).

So the first, and main, problem with stakeholder mapping is getting to grips with being really clear exactly who are your stakeholders in any given venture, enterprise, or business. Specification is all. The trick is to avoid using generic terms for groups involved and/or affected by the organization or enterprise, and to focus on its specific inputs, processes, outputs, and outcomes.

Stakeholder mapping is a management tool of key value to any innovator at any and all stages on the 10-step innovation process set out in the Introduction (ref. pp. 24–25 above). It applies and has value in business, public/civic, and not-for-profit/voluntary sector organizations, but in different ways that reflect the differences between these three sectors. Business organizations will be focused on their customers and clients, and may include these as parts of their list of company stakeholders, along with shareholders, suppliers, business partners, and those communities in which their businesses operate (as incorporated into international and local excellence frameworks such as EFQM as Society Results). The trick for innovators in business organizations is therefore to identify quite precisely and specifically who the organizations' various stakeholders are, as separate and distinct from customers and clients who make the firm its money.

In doing this, business organizations can look at how both public sector and not-for-profit/voluntary organizations view their stakeholders. For these organizations, even the idea of *customers* (as in, paying customers with real choices available) can be contested. In such a business and operating environment, the value of identifying stakeholders can help an organization and its key players (management, top executive, and board) to anticipate responses and reactions from different groups within and outside the organization.

The main uses of stakeholder mapping include

- Competitive positioning in the market, for example, as part of cost–benefit analysis of possible ventures
- Community/local impact assessment of plans and proposals for business operations (e.g., location of plants, offices, outlets, distribution, logistics in support of new ventures)
- Pre- and postevaluation of individual innovations in products, processes, services, and programs in terms of community social and environmental impacts
- External-facing corporate social responsibility and corporate citizen programs and activities of the business/organization
- Internal corporate assessments of impacts and responses to change programs and business/organizational innovations
- As a tool for management and boards in assessing power and influence, for example, in sponsoring innovation—both "Big I" (high-risk) and "Little I" (lower-risk) ventures

The main types of stakeholder mapping across business, public, and not-for-profit/voluntary sectors include

- Mapping relationships and interests of key players, identifying key stakeholders, and mapping degrees of support (+) or opposition (−) to particular plans and proposals. This can be shown in matrix form, as in the power–interest quadrant matrix of Scoles and Johnson. More simply, this can be illustrated as key stakeholders sitting round the table, as if at an executive or board meeting to make a key decision on plans and proposals.
- Mapping the key processes and steps in developing corporate communications strategies to build and maintain stakeholder relationships, usually for products, processes, and services being marketed and delivered for a reasonable period of time, and for whom sets of stakeholders can be identified besides the customer segments and groups being aimed at.

As a general rule, both types of stakeholder mapping are more effective when innovators focus on the few key influencers and decision makers, at each stage of the innovative process.

Making Use of the Tool

Power–Interest Quadrant Matrix

As developed by Mendelow, Johnson and Scholes, Darwin and others in Johnson and Scholes Exploring Corporate Strategy (2008), the matrix shows the power of various stakeholders (high or low) plotted against the level of interest of these stakeholders (high or low).

Creating a stakeholder map involves the following steps:

1. Deciding which stakeholders to plot
2. Assessing power, referring to sources and bases of power for each identified stakeholder
3. Assessing interest
4. From the matrix, establishing where things lie in terms of likely support or opposition, and looking to see where any gaps or mismatches can be sorted to promote support for the innovation

None of the above implies that it is ethical or effective to adopt Machiavellian schemes of manipulation or control. Rather, it is simply a matter of prudent planning and common sense to check out the lay of the land, understand where support or opposition may come from, then plan and respond accordingly. It is indeed *political* behavior, and considering such factors is a necessary condition for sustained success for any innovation, not only in the early innovative phases but also in the postproduction performance analysis and *selling* of the innovation story. Too many businesses and organizations have failed to reap sustained success for their innovations, letting them go to others for effective exploitation and development.

Communications Tracking Stakeholder Mapping

In the EFQM/Speller Communications Tracking model for stakeholder mapping, there are seven key stages for organizations to consider:

1. Identify key stakeholders
2. Identify the nature of impacts for each key stakeholder group
3. Identify the nature of the relationship and communication that works for each key stakeholder group
4. Develop the mapping of communications and consultation processes specific for each group
5. Identify and develop tracking methods of stakeholder perceptions
6. Monitoring and reporting of key measures of stakeholder perceptions
7. Follow-up action on stakeholder feedback, and monitoring and reporting on this
8. And of course, Deming style, do it all over again

The key to effective support of innovation in this approach is keeping an eye continuously on one's stakeholders, as much as one should do in customer feedback tracking. And in the case of public sector and not-for-profit/voluntary innovation, it is most likely that stakeholder perceptions of impacts can play as big a role in successful sponsorship of innovation as actual impacts.

EXAMPLE

In this example (see Figure 13.1) of the power–interest quadrant matrix, four sets of stakeholders are identified in a scenario where new management systems are being introduced to improve service quality and performance management in a municipal authority we can call Strongoak,

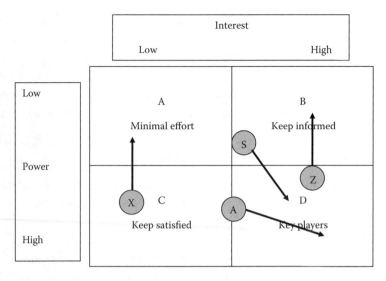

FIGURE 13.1
Power–interest quadrant matrix.

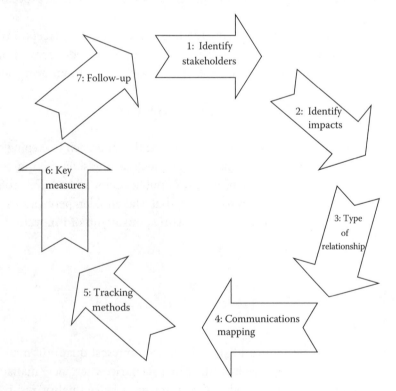

FIGURE 13.2
Communications tracking stakeholder mapping.

in the United Kingdom. What is happening is that traditional interests such as the constituency political party in power (X) is being increasingly marginalized, executive councilors (S) are becoming more managerial in approach to innovation and change management, yet are themselves less important than the political group leader/executive mayor (A) who is becoming more autocratic and through whom all major decisions are made. Her predecessor was content with episodic interventions, and to leave professionals and officials to run the show day to day. In the meantime, those elected councilors not on the executive (Z) are being marginalized. They are well informed but have no real say in what goes on, least of all in important innovations in services design and delivery.

This seven-step tool allows the innovator to check at each step (Figure 13.2)

- *What* is needed to be known or done at each step
- *Who* is expected to lead
- *How* communications and consultations should be carried out, both directly and indirectly

A full account is set out in Foley et al. (1997). Briefly, each step is outlined below:

1. Identify your key organizational stakeholders: Very much a job for leaders and top managers, as well as those driving innovations, and using contact staff as eyes and ears about impacts and perceptions. Formal and structured feedback and informal feedback can help.
2. Identify the impacts for each key stakeholder group: The key here is to be as specific as possible, to distinguish different possible impacts for different groups—to assess depth and intensity of impacts, and frequency/time factors.
3. Analyze relationship of organization with stakeholders: Building on steps 1 and 2, this step is about ensuring nothing is being missed, and that a clear picture of what is salient and appropriate for each stakeholder group is being understood. This is about effective listening, and recognizing different tailored approaches work with different groups.
4. Communications mapping: This step is about process mapping communications and consultations at each phase of the innovative cycle. This really has to be SMART.

5. Tracking methods: Having done the mapping, this step is about tracking perceptions and impacts. This may best be done by marketing and sales professionals. In a small innovation or boutique start-up, this is down to the innovator. Making tracking a good habit is key.

6. Key measures: As things develop and more is known about which methods and measures truly give useful feedback, this step is about refining those key measures. What are the *critical few* factors (e.g., using 80:20 Pareto analysis)?

7. Follow-up: "Do it all over again." Innovation is relentlessly knowledge driven.

SOFTWARE

No specific software is recommended.

REFERENCES

Burlton, R. *Business Process Management*. Indianapolis, IN: SAMS, 2001.

Foley, G., Speller, S., and Boyce, J. *Impact on Society—Opportunity for Action*. EFQM, 1997.

Johnson, G. and Scholes, K., eds. *Exploring Public Sector Strategy*. Harlow, UK: Prentice Hall, 2001.

Johnson, G., Scholes, K., and Whittington, R., eds. *Exploring Corporate Strategy*, 8th Ed. Harlow, UK: Pearson Education, 2008.

14

Innovation Master Plan Framework

Langdon Morris

CONTENTS

DEFINITION

The innovation master plan framework consists of five major elements. They are strategy, portfolio, processes, culture, and infrastructure.

USER

This tool is normally used to create the master plan by the executive committee. In some cases, a subcommittee is set up to prepare a rough draft that is approved by the executive committee. Lower-level detailed plans to support the master plan and apply it to the individual functions are normally assigned to middle and first-line managers, with the executive team approving their plans. Everyone within the organization should be affected with the way the innovation master plan is implemented.

OFTEN USED IN THE FOLLOWING PHASES OF THE INNOVATIVE PROCESS

The following are the seven phases of the innovative cycle. An X after the phase name indicates that the tool/methodology is used during that specific phase.

- Creation phase X
- Value proposition phase
- Resourcing phase
- Documentation phase
- Production phase
- Sales/delivery phase
- Performance analysis phase

TOOL ACTIVITY BY PHASE

The innovation master plan is an overview document that is used to guide all of the projects, programs, and activities within the organization. Its output applies to all phases that make up the innovation cycle. It is the primary document that defines the difference between if an idea/concept is just creative or if it is innovative. Innovative ideas are in line with the business plan and can create value for the organization and its customer.

HOW TO CREATE AND USE
THE INNOVATION MASTER PLAN

Innovation is vitally important, as we all know that businesses and governments must innovate to survive. But it remains very difficult for most organizations to achieve innovation on a consistent basis. We only have to look at the recent history of global businesses to see the impact of innovation—new and innovative companies are achieving great success, while the companies and even nations that do not innovate often fail and fall behind.

In practice, it is obvious that a significant part of the innovation process depends on the creative capacity to come up with new and compelling ideas, while another aspect of success at innovation has to do with how we organize and conduct the innovation process from a technical and managerial perspective. The intent of the master plan framework is to address the technical and managerial elements, so the focus here is on the principles, tools, and methods necessary to a systematic and rigorous business process that can achieve meaningful and long-lasting innovation results.

The innovation master plan framework consists of five major elements, which are associated with five key questions—why, what, how, who, and where—and therefore this summary is organized into these five topics.

Strategy

Why Innovate: Link between Strategy and Innovation

The *why* of innovation is simple: change is accelerating, and we do not know what is coming in the future, which means that we must innovate to both prepare for change and to make change. Hence, the why of innovation is brutally simple: change is accelerating.

If things did not change, then your organization could keep on doing what it has always done, and there would be no need for innovation. If markets were stable, if customers were predictable, if competitors did not come up with new products and services, and if technology stayed constant, then we could all just keep going as we did yesterday.

But all the evidence shows that change is racing faster and faster, which means many new types of vulnerabilities. Technology advances relentlessly, altering the rules of business in all the markets that it touches, which is of course every market. Markets are not stable, customers are completely fickle, and competitors are aggressively targeting your share of the pie. So please ask yourself, "Are we managing with the realities of change in mind? And are we handing uncertainty?"

Since the alternatives are either to *make change* or to *be changed*, and making change brings considerable advantages while being changed carries a truckload of negative consequences, then the choice is not really much of a choice at all. You have got to pursue innovation, and you have got to do it to obtain long-lasting benefits.

The decisions to be made focus on how best to prepare for future markets, and the actions relate to transforming the innovation mindset into meaningful work throughout the organization, work that results in the development of innovations that influence the market, and improve the position of the organization relative to its competitors.

Hence, innovation plays a critical role for many firms. Do you admire Apple or Google? Then ask yourself what role innovation plays in Google's strategy. It is obvious that we would not admire Apple or Google, and in fact we would not even know about either one, if it were not for innovation. The very existence of these companies is based on strategic insight and on critical innovations that made the strategy real. For Google, the critical insight was that as the number of web pages grew, the Internet's potential as an information resource was surpassing all other resources for scale, speed, and convenience; however, it was getting progressively more difficult for people to find the information they were looking for.

People therefore came to value better search results, and Google's first innovation to address that need was its PageRank system, developed in 1995, an algorithm for Internet searches that returned better results than any other search engine at the time.

Google's second innovation was a business model innovation, which turned the company into a financial success along with its technical search success. When Google's leaders realized in 2000 that they could sell advertising space at auction in conjunction with key words that Google users searched for, they unleashed a multibillion dollar profit machine. The integration of these two innovations provided a multiplicative advantage, and Google's competitors are falling by the wayside as the company continues to dominate.

What other companies do you like? Do you also admire Starbucks? Or Disney? Or Toyota? Or BMW? They are certainly innovators, and many of us appreciate them precisely because of it. So the relationship between strategy and innovation is vital, and the important role that innovation plays in transforming the concepts of strategy into realities in the market-place tells us that none of these companies could have succeeded without innovation. This is the *why* of innovation.

Portfolio

What to Innovate: Creating and Managing Innovation Portfolios

Investors in all types of assets create portfolios to help them attain optimal returns while choosing the right level of risk, and innovation managers must do the same for the projects they are working on.

Innovation is inherently risky. You invest money and time, possibly a lot of both, to create, explore, and develop new ideas into innovations, but regardless of how good you are, many of the resulting outputs will never earn a dime.

Is that failure or success? It could be both. The degree of failure or success will be determined not by the fate of individual ideas and projects, but by the overall success of all projects taken together. Hence, the best way to manage the risk is to create an innovation portfolio.

Just as investors in all types of assets create portfolios to help them attain optimal returns while choosing the level of risk that is most appropriate for them, you will do the same for the innovation projects you are working on by allocating capital across a range of investments to obtain the best return while reducing risk.

The underlying principle of portfolio management is that the degree of risk and the potential rewards have to be considered together. In a rapidly changing market, the nature of innovation risk is inherently different than in a slower-changing industry such as, say, road construction, because the faster the rate of change in a company's markets, the bigger the strategic risks it faces. The faster the change, the more rapidly will existing products and services become obsolete, a factor we refer to as *the burn down rate*. The faster the burn down, the more urgent is the innovation requirement.

This will necessarily affect the composition of an innovation portfolio by inducing a company to take greater risks in innovation its efforts. Hence, the ideal innovation portfolio of each organization will necessarily be different: Apple, NASA, Genentech, Union Pacific, GE, and Starbucks are all innovative organizations; however, when it comes to their innovation portfolios, it is obvious that they cannot be the same in content or style.

A further key to the dynamics of a successful portfolio is described in portfolio theory, which tells us that the components of a portfolio must be noncorrelated, meaning that various investments need to perform differently under a given set of economic or business conditions. In the case of innovation, *noncorrelated* means that every firm needs to be working on potential innovations that address a wide range of future market possibilities in order to assure that the available options—and here is the key point—will be useful under a wide variety of possible future conditions.

The need for broad diversity in the portfolio also reminds us we need to develop all four types of innovation, so what we are really talking about are five different portfolios. There will be a different portfolio for each type of innovation, breakthroughs, incremental innovations, new business models, and new ventures, and there will be a fifth portfolio that is an aggregate of all four.

We should also note that each different type of portfolio will be managed in a different process, by different people, who have different business goals, and who are measured and possibly rewarded differently. Hence, metrics and rewards are inherent in the concept of the portfolio, and the master plan also calls for the design of the ideal metrics by which the portfolio should be measured.

And because we are preparing for a variety of future conditions, it is obvious that some of the projects will never actually become relevant to the market, and they will therefore never return value in and of themselves. But this does not mean that they are failures; it means that we prepared for a wide range of eventualities, and some of those futures never appeared, but we were nevertheless wise to prepare in this way. This sort of *failure* is a positive enhancement of the likelihood of our survival and ultimate success, so it is not failure in a negative sense at all. By analogy, I carry a space tire in my car, but it is not a failure if I never have occasion to use it.

Therefore, the process of creating and managing innovation portfolios cannot be managed by the chief financial officer's (CFO's) office as a purely financial matter. Instead, the finance office and innovation managers are partners in the process of innovation development. Hence, innovation portfolio management is like venture capital investing, early-stage investing where it is impossible to precisely predict the winners, but nevertheless a few great successes more than make up for the many failures.

And the CFO will also have to accept the idea that the mandatory investments in innovation mean investing in learning. During the early stages of the development of an idea, its future value is almost entirely a matter of speculation. As work is done to refine ideas in pursuit of business value, the key to success is learning, as the learning shapes the myriad design decisions that are inevitably needed. The innovation process as a whole therefore seeks to optimize the learning that is achieved, and to capture what has been learned for the benefit of the overall innovation process as well as the portfolio management process. This costs money, which cannot and should not be avoided.

As the projects that constitute an innovation portfolio mature and develop, they provide senior executives and board-level directors with increasingly attractive new investment options. Furthermore, by managing their portfolios over time, a team of executives can significantly improve the portfolio's performance, for as they engage in this type of thinking they get more in sync with the evolving market, and often better at identifying and supporting the projects that have greatest potential.

Still, many will fail. In fact, a healthy percentage of projects *should* fail, because failure is an indication that we are pushing the limits of our current understanding hard enough to be sure that we are extracting every last bit of value from every situation, and at the same time preparing for a broad range of unanticipated futures.

Process

How to Innovate: Innovation Process

Many people assume that creating new ideas is the beginning of the innovation process, but actually that is not true. Ideation occurs in the middle of the disciplined innovation process.

While the purpose of innovation is *simply* to create business value (*simply* is emphasized because it is obviously not so easy to do), the value itself can take many different forms. As we noted above, it can be incremental improvements to existing products, the creation of breakthroughs such as entirely new products and services, cost reductions, efficiency improvements, new business models, new ventures, and countless other forms as well.

Overall, the method of creating innovation is to discover, create, and develop ideas, and to refine them into useful forms, and to use them to earn profits, increase efficiency, or reduce costs. So in the quest for innovation, it is obvious that many ideas at the input stage become a few completed, useful innovations at the output stage. This is why people readily visualize the innovation process as a funnel: lots of ideas come in the wide end on the left, and a few finished innovations come to market from the narrow end at the right. The trick to making it work is to know what is supposed to happen inside the funnel.

So, naturally, you want to start by creating a whole bunch of ideas, right? Actually, the answer is no—ideas are indeed the seeds of innovation, just as ore taken from the ground is the raw material of steel, or waving fields of wheat provide the raw material for bread. But it takes a lot of work to mine the raw ore and transform it into steel, or to prepare the fields to grow the wheat long before it becomes bread. It is the same with innovation; we do not start by collecting raw ideas. Instead, we know that innovation is a core element of our organization's strategy, so we have to start the innovation process itself with strategic thinking to assure that the outputs of innovation are fully aligned with our strategic intent.

Step 1 is therefore *strategic thinking*. The innovation process begins with the goal to create strategic advantage in the marketplace, so in this stage we think specifically about how innovation is going to add value to your strategic intents, and we target the areas where innovation has the greatest potential to provide strategic advantage.

Step 2 is *portfolio management and metrics*. As we discovered above, one of the important underlying facts of innovation management is the necessity of failure. We are, by definition, trying to do something new, and as we proceed on the innovation journey we do not in fact know if we are going to succeed. We have confidence that we will succeed eventually, but along the way we know that there will be many wrong turns, and many attempts that will never come to fruition. So we manage innovation portfolios aggressively to balance the inherent risks of the unknown with the targeted rewards of success, and balancing our pursuit of the ideal with the realities of learning, risking, and failing in order to ultimately succeed.

Steps 1 and 2 together provide a platform and context for everything that follows, and so they constitute the *input* stages of the funnel, and so that the activities in stages 3 through 7 have the best chance to achieve the best results.

Step 3 is *research*. An output of stage 2 is the design of the ideal innovation portfolio, which is what we believe, as of today, is the right mixture of short- and long-term projects across all four types of innovation. Once we understand the ideal, we can compare our current knowledge and discern the gaps. Filling these gaps, then, is the purpose of research. Through research, we will master a wide range of unknowns, including emerging technologies, societal change, and customer values, and in the process we will expose significant new opportunities for innovation (see Figure 14.1).

Strategic thinking has clarified for us how the world is changing and what our customers may value, and this stimulates new questions that our research has answered. Research findings provoke a broad range of new ideas across a wide range of internal and external topics. This is the abundant raw material, and it is already and automatically aligned with our strategic intent because it came about as a result of a direct connection between strategy, portfolio design, and research.

Step 4 is *insight*. In the course of our explorations, the light bulb occasionally illuminates, and we grasp the very best ways to address a future possibility. Eureka! The innovation and the target and mutually clarified; we understand what the right value proposition is for the right customer.

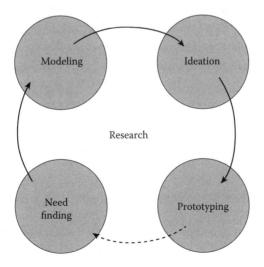

FIGURE 14.1
Four stages of the research cycle.

So while many people think of this moment of insight as the beginning of the innovation process, as you can see, in the well-managed innovation effort, we expect insight to come about as the result of the preceding processes and activities, not at random.

Hence, the innovation process described here is specifically contrasted with random idea generation: insight is the result of a dedicated process of examination and development. It does not occur because someone had a good idea in the shower, but because individuals and teams of people were looking diligently and persistently for it.

Step 5 is *innovation development*, the process of design, engineering, prototyping, and testing that results in finished product, service, and business designs. Manufacturing, distribution, branding, marketing, and sales are also designed at this step in an integrated, multidisciplinary process.

Step 6 is *market development*, the universal business planning process that begins with brand identification and development, continues through the preparation of customers to understand and choose this innovation, and leads to rapid sales growth.

Step 7 is *selling*, where the real payoff is achieved. Now we earn the financial return by successfully selling the new products and services. In the case of process improvement innovations directed internally, we now reap the benefit of increased efficiency and productivity.

Managing a process of this scope and complexity is of course a challenge for all organizations, but among the world's companies we see that there are some that do this extraordinarily well. The knowledge that some do it very well, and that it is certainly possible to be an exemplary innovative organization that can attain exceptional profits, should be a powerful source of motivation to develop and apply your own master plan.

Culture

Who Innovates: Creating the Innovation Culture with Geniuses, Leaders, and Managers

Organizations that are successful at innovation naturally develop a strong innovation culture. Such a culture is much appreciated by customers who say that the company is a genuine innovator, and it is also known among the people inside the organization as a dynamic and innovation-friendly place to be.

But supposing an innovation culture does not yet exist in your organization. Then, how can you nurture it? How do organizations develop an innovation culture? Who should be involved in the innovation process? And what roles should they play?

Every culture is an expression of behaviors and attitudes, and every organization's culture reflects the beliefs and actions of its people, as well as the history that shaped them. The innovation culture, of course, is likewise an expression of people, their past, and their current beliefs, ideas, behaviors, and actions about innovation. We have found that the innovation culture comes into being when people throughout the organization actively engage in promoting and supporting innovation, implementing rigorous innovation methods, and filling three essential roles: creative geniuses, innovation champions, and innovation leaders.

Innovation's Creative Geniuses

Who comes up with the critical ideas that are the beginnings of innovation, and then turns these ideas into insights, and insights into innovations? They are creative geniuses, and they work everywhere, inside and outside.

If it seems like a stretch to label these people as *geniuses*, let me explain the rationale. No one can innovate if they accept things the way they are today, so making innovations requires that we are willing to see things differently. We have to overcome institutional and bureaucratic inertia that may burden our thinking process, and challenge ourselves to see beyond conventional viewpoints. This fits perfectly with the dictionary definition of genius, which is "exceptional natural capacity shown in creative and original work."

Innovation Champions

Innovation champions (also referred to as *innovation managers*) are those who promote, encourage, prod, support, and drive innovation in their organizations. They do this in spontaneous moments of insight, in ad hoc initiatives, as well as in highly structured innovation programs.

Innovation champions build the practical means for effective, systematic innovation. They take direct responsibility for finding creative thinkers and encouraging them to see and work in new ways; they help people seek new experiences that may spark new ideas; and they create a regular operations context in which sharing and developing new ideas is the norm.

While they may work anywhere in the organization, including in senior management positions, line management roles, staff, or frontline operation roles, the specific nature of the innovation champion's role is to function in the middle, to provide the bridge between the strategic decisions of senior managers, and the day-to-day focus of frontline workers.

Innovation Leaders

An innovation leader is someone who shapes or influences the core structures and the basic operations of an organization, all with a clear focus on supporting innovation. Core structures include the design of the organization itself, as well as its policies and their underlying principles. Metrics and rewards can also be core structures.

None of these factors are absolute givens, and all of them can be changed, and that is the point: they are all subject to design, to thoughtful choice about what is best. It is generally within the power of senior managers to change them, and when they impede innovation they should be changed to favor it.

The actions and attitudes of senior managers are based, ultimately, on their philosophies about management, on their mindset, which we explored earlier in this document. Innovation leaders set expectations, define priorities, celebrate and reward successes, and deal with failures, and all of these factors can be done in a way that makes innovation easier or more difficult, because each can be arranged to favor the status quo or to favor useful and effective change.

Do leaders believe in a win–win model, or win–lose? Win–lose organizations usually are not trusting environments, and because trust is so important to innovation, when it is missing, innovation suffers.

Leaders also set goals, and they do not need to be modest; in fact, they can be outright aggressive. By setting ambitious goals, managers emphasize the linkage between an organization's strategy and the pursuit of innovation, elevating innovation to a strategic concern where it properly belongs. Conversely, if innovation is not expressed as a specific goal of top management, then it probably would not be a goal of anyone else, either; and if policies are restrictive and make it difficult to test new ideas, then there would not be many new ideas. We refer to organizations that are focused on the present, rather than the future, as *status quo organizations*.

The firm that is obsessed with the status quo probably would not last very long, but some managers still seem to believe in this model, and their domineering attitudes and behaviors reinforce it.

Innovation does not happen without leaders who embrace it, nor can it happen without people who have ideas and are willing to risk failure to experiment with them. Nor does it happen without champions to bridge between the strategic and operations questions and the individuals who have ideas and want to explore them.

And of course it happens best, and fastest, when all three roles are consciously implemented and mutually supporting. This does not mean that each individual can play only one of these roles; many people are geniuses, and leaders, and champions, and at various times we play all of these roles.

So what is important is not that we classify people into the various categories; in fact, we should avoid doing that. We just need make sure that all three roles are being played, and played well, so that defining, developing, and implementing ideas that become innovations becomes the norm.

Infrastructure

Where We Innovate

Organizations that consistently deliver innovation do so because their employees have the skills to effectively explore, understand, diagnose, analyze, model, create, invent, solve, communicate, and implement concepts, ideas, and insights. These are all attributes that we might consider facets of *learning*, and naturally enough any organization that thrives in a rapidly changing environment necessarily has developed the capability to learn and to apply that learning to keep up with external changes.

Certainly, the link between learning and innovation is a strong one, and clearly speed matters. The faster people in a company can learn, the faster they can apply that learning to create the next product, service, and business model. By creating a positive and self-reinforcing feedback loop of accelerated learning to create innovation, organizations then obtain more learning, leading to more innovation. The results are manifold: shorter product life cycles, which leads to quicker learning, better profits, etc., all contributing to competitive advantage.

To support the acceleration of learning and innovation, we have found that the proper infrastructure tools make a big difference. The four key infrastructure elements are open innovation, effective collaboration, the virtual workplace, and the design of the physical work place.

- Open innovation
  While in the past, many organizations kept the innovation process closely guarded as an in-house secret, these same companies have recently discovered that seeking new product ideas from outside

can significantly improve the flow of new opportunities. Applying the principles of open innovation can significantly accelerate the pace of innovation, as well as its effectiveness. Open innovation means expanding the pool of participants in the innovation process to all types of outsiders, including customers, suppliers, partners, and community members, tapping into ideas, critical thinking, and advice.

- Collaboration

  Everyone who works in the field of innovation agrees that collaboration is vital to success at innovation. Mastering and applying the principles of effective collaboration, not only for pairs and small groups, but also for groups of tens or even hundreds of people, requires facilitation skills to help nurture new ideas and turn them into effective innovation, and the benefits can be significant.

- Virtual work place

  As we spend more and more time working and collaborating online with our internal colleagues and with outside partners, customers, and vendors, the quality of our tools and our skill in using them can make a significant difference in the productivity of our innovation efforts. Active engagement in the selection and adoption of the right tools is a simple but fundamental rule to follow.

- Physical work place

  As Massachusetts Institute of Technology Professor Tom Allen puts it in the lively book he co-authored with architect Gunter Henn, called *The Organization and Architecture of Innovation*, "Most managers will likely acknowledge the critical role played by organizational structure in the innovation process, but few understand that physical space is equally important. It has tremendous influence on how and where communication takes place, on the quality of that communication, and on the movements—and hence, all interactions—of people within an organization. In fact, some of the most prevalent design elements of buildings nearly shut down the opportunities for the organizations that work within their walls to thrive and innovate. Hence, the implications of physical space for the innovation process are profound."

The essentials for effective innovation are thinking, creating, problem solving, and collaborating, and we know that the work place that best supports them is not a traditional conference room, but a much better work environment that is designed for innovation.

These four elements, open innovation, collaboration, the virtual workplace, and the physical workplace, constitute the critical elements of the innovation infrastructure, and it is by providing these tools to the innovative people in your organization that you can help them do their best to develop the innovations that will compose your organization's future.

Summary of Technical and Cultural Factors to Consider

The innovation master plan covers each of these five of the major elements in great depth. A comprehensive review of your organization's performance as an innovator will clearly identify what is working well and what is not working at all, and to design the corrections. We have found that seven technical factors are critical to innovation performance, and seven additional key factors are cultural.

Seven technical factors: We refer to these as technical because they can be assessed in a relatively objective fashion, and according to specific technical criteria.

- Alignment of strategy and innovation
  - You will describe how you intend to align the innovation process with the strategy process.
- Innovation portfolio management
  - You will assess the contents of the existing innovation portfolios to see if they have the right balance of incremental and breakthrough projects, and to assess the projects that are under way to determine if they really are the right future products and services for the organization to make and sell.
- Research
  - The purpose of assessing the research process is to determine how well it is capturing the critical tacit knowledge that will feed the search for unknown and unmet needs.
- Innovation development
  - You will evaluate the development process as well, to make sure that innovation development and market development are proceeding effectively and in parallel, and providing the right guidance for the organization.

- Alignment with sales
    - You will make sure that the new products and services that are being introduced are effectively aligned with the sales organization so that the organization can, in fact, bring these products and services to market effectively.
- Innovation metrics and rewards
    - You will determine which innovation metrics should be used throughout each stage of the innovation process, and make sure that those innovation metrics are aligned with the rewards that are offered to individuals, teams, departments, and business units.
- Infrastructure
    - And you will examine the infrastructure to determine whether the people who are working in the innovation process have sufficient information and support to complete their work as efficiently as possible.
    - Wherever a stage of the process or a critical skill is not at the level it ought to be, or if it is missing entirely, you will design an improvement plan or a process to implement it from scratch.

Cultural factors: These factors are perhaps more subjective than the technical factors, but they are nevertheless critical to effective innovation performance as well.

- Innovation culture
    - A critical issue, of course, is the character of the organization's culture. Does it favor innovation, or shun it? Do people feel safe in taking risks, or is this a career-threatening move, and something that is consistently avoided? Do people embrace the characteristics and qualities of the innovation culture, or is it an organization that seeks the status quo culture?
- Creativity
    - You will assess the creativity of people throughout the organization to see if the new ideas that are under development are sufficiently creative.
- Trust
    - You will examine the level of trust in the organization to see if people are comfortable in their working relationships to allow the ambiguities and uncertainties of the innovation process to

follow a natural developmental flow, or if the lack of trust forces people to make innovation decisions too quickly because it is not safe to allow ambiguity to resolve itself over time.

- Leadership
  - You will assess the performance of leadership across the four delivery factors, including goal setting, expectation setting, and tone setting.
- Mindset
  - You will assess the mindset of the leadership team to see how well they understand the acceleration of change, and how much they are prepared to support the innovation process as the development of the future products and services for the company.
- Attitude
  - You will review the overall attitude that people have toward innovation to find out if innovation is sufficiently support, and if people are engaged in the innovation process.
- Tone
  - And, of course, you will look at the tone as defined as said by senior leadership to see that innovation is getting the proper support in both words and actions.

A thorough assessment involves talking with a lot of people, and not just people in top management, but people throughout the organization. We may also interview people who are outside, including customers, suppliers, and partners, to learn their views on our organization's innovation performance, strengths, and weaknesses. Interviews often last 30 to 60 minutes, although sometime these last longer than that.

Researching the external environment measures the rate of change in the market, and assesses the innovation capabilities and performance of major competitors.

To learn the views and experiences of a larger group of people, it is helpful to do an online survey to reach hundreds more people. Ten to twenty minutes of questions and answers, provided anonymously and therefore candidly, provide tremendous depth of information about people's attitudes, feelings, and experiences of the innovation process.

By comparing findings across all three of these information sources, we expect to gain a detailed understanding of current innovation performance,

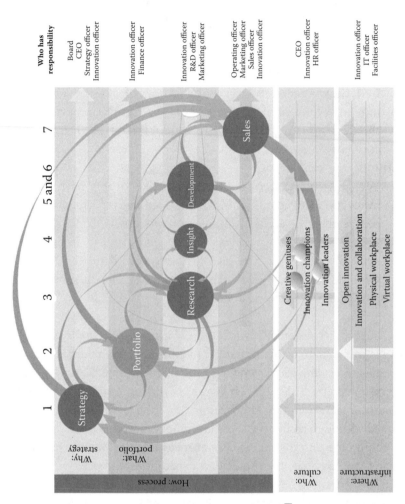

FIGURE 14.2

Complete framework for the master plan.

assess where performance is outstanding, where it is good, and where it needs to be improved, and where it should be targeted.

The findings will be written up into a plan, and may also be the subject of a briefing with the executive team to help them understand the organization's current capabilities, prescriptions, and the role that each member of the executive team needs to play in developing and promoting the dramatically enhanced innovation capability that you envision, the transformation of your organization into a genuine innovator.

As you prepare the innovation master plan, also expect to identify 10 to 20 major improvement areas to focus on, and these may become the primary drivers of your innovation action plan for the first 6 months to 1 year. During the course of that work, you will be interacting with a great many people, and both coaching and encouraging them to participate in the innovation process, as well as for many insisting that they follow the rigorous innovation structure.

While innovation is a challenge for most organizations to achieve, it is also fun, fascinating, and *very* rewarding. There are few accomplishments as satisfying as seeing your ideas and decisions make a positive difference for your organization.

Figure 14.2 provides an excellent picture related to the activities reported in this chapter.

EXAMPLES

Examples have been discussed in the section entitled "How to Create and Use the Innovation Master Plan."

SOFTWARE

- SmartDraw: smartdraw.com
- Asana Guide: asana.com
- i-Nexus: Sales@i-nexus.com

SUGGESTED ADDITIONAL READING

Harrington, H.J. and Voehl, F. *The Organizational Master Plan Handbook—A Catalyst for Performance Planning and Results*. Boca Raton, FL: CRC Press, 2012.
Morris, L. *The Innovation Master Plan*. Walnut Creek, CA: Innovation Academy, 2011.

15

Knowledge Management System

Achmad Rundi

CONTENTS

Knowledge will light the hallway to success.

H. James Harrington

DEFINITION

Knowledge management (KM) is a strategy that turns an organization's intellectual assets, both recorded information and the talents of its members, into greater productivity, new value, and increased competitiveness. It is the leveraging of collective wisdom to increase responsiveness and innovation.

USER

It often is a database controlled by an individual or group that makes the information available to a community of practice.

OFTEN USED IN THE FOLLOWING PHASES OF THE INNOVATIVE PROCESS

The following are the seven phases of the innovative cycle. An X after the phase name indicates that the tool/methodology is used during that specific phase.

- Creation phase X
- Value proposition phase X
- Resourcing phase X
- Documentation phase X
- Production phase X
- Sales/delivery phase X
- Performance analysis phase X

TOOL ACTIVITY BY PHASE

People will create and utilize knowledge at many stages of innovation. Knowledge output from earlier stage can be utilized in the next stage and that can generate next knowledge, transferrable to the next stage of innovation.

Nonaka and Takeuchi (1995) see this new knowledge dissemination as leading to innovation that is systematic and continuous.

McAdam, 2000

HOW THE TOOL IS USED

What is knowledge management (KM)? Knowledge management delivers value from the knowledge that a group/organization/firm wants to create and utilize in order to sustain its business through a systematic and cyclical process. Gartner Group has expanded the definition of KM concept as follows:

> Knowledge management is a discipline that promotes an integrated approach to identifying, capturing, evaluating, retrieving, and sharing all of an enterprise's information assets. These assets may include databases, documents, policies, procedures, and previously uncaptured expertise and experience in individual workers.
>
> **Koenig, 2012**

KM was first heard of or implemented in the consulting community (Koenig, 2012). With the enablers such as Internet and Intranet, they linked their geographically dispersed knowledge-based organizations, and exchanged knowledge. The principles of such KM were then adapted by many consulting organizations and also other disciplines. As the consulting community gained expertise to link and manage knowledge through Intranets, they understood they have a service or a product that would benefit and be sellable to other organizations. They came up with *knowledge management* as the name. Over the course of time, KM has evolved in three phases (Koenig, 2012), and interest in KM has grown steadily as shown by accumulating articles on KM over the years.

We have since moved on from neoclassical economy to knowledge economy. Knowledge is not just a mere collection of data. It is also what we have learned during the business activities and also kept to ourselves after we succeeded or failed at what we are doing. People need to share knowledge and build on it to address the objectives. We have then treated knowledge better as assets that we can trade because of the opportunity it can bring. Peter Drucker said that the only meaningful resource is knowledge (Nonaka and Toyoma, 2002).

To really understand KM, we need to know also the dynamic or essence of knowledge. This is based on Nonaka's and Toyoma's work to refine the definition of firm (organization) from the knowledge perspective. Nonaka and Toyoma (2002) refined the definition of a *firm* as the production unit that needs to look more from the perspective of knowledge instead of just functions that feed output of tasks to the following tasks. The firm also takes into consideration that the knowledge has cost. Knowledge has to be created and utilized. It is also needs to be converted from tacit knowledge to explicit knowledge to be useful to many.

- Tacit (soft) knowledge is defined as knowledge that is framed around intangible factors embedded in an individual's experience. It often takes the form of beliefs, values, principles, and morals. It guides the individual's actions. It is embedded in the individual's ideas, insight, value, and judgment. It is only accessible through the direct corroboration and communication with the individuals that have the knowledge.
- Explicit (hard) knowledge is knowledge that is stored in a semistructured content, such as documents, e-mail, voicemail, or video media. It can be articulated in formal language and readily transmitted to other people. Also called hard knowledge or tangible knowledge, it is conveyed from one person to another in a systematic way.

The knowledge has both interaction and justification cost related to obtaining and using it. (Nonaka and Toyoma, 2002). The first one is due to the nature of knowledge creation that happens through interaction of individuals or units from different contexts. The more common the context that has been established already, the less cost, which can happen internally within the firm's shared context. The second cost is justification because in order for the knowledge to be utilized, it has to justify the truth of the knowledge, whether it was created internally or obtained from external market.

As mentioned above, the change of tacit knowledge to explicit knowledge goes through modes of conversion. It is the socialization, externalization, combination, internalization (SECI) process where knowledge goes into spiral and continuous knowledge creation and utilization, and expands in quality and quantity from individual to the group, and then to the organization. Figure 15.1 is the diagram of the SECI process.

The value of the knowledge will also be more useful when the *firm* can synthesize contradictions in its business, such as issue of globalization and localization, and other opposites, then find the balance (Nonaka and Toyoma, 2002). It can be dynamic over time where the contradictions change. The firm needs to take time to build its tacit knowledge assets.

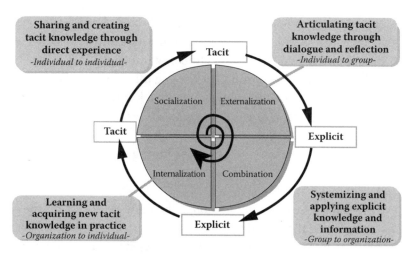

FIGURE 15.1
Diagram of SECI process.

Then, it will have to be quick to convert the necessary tacit knowledge into explicit knowledge that will be useful to address it.

The better the firm at managing its knowledge, especially the knowledge exchanged in the market, the better it is at sustaining itself and having competitive advantage.

KM is not solely on the managing the knowledge itself. KM will enable the firm/organization to provide the necessary knowledge to individuals in order to provide more value to the organization. KM will help organizations use knowledge effectively to meet their needs (Harrington, 2011, p. 155). The focus of the KM will be to understand how the individuals acquire knowledge and what additional knowledge they will need to add more value (Harrington, 2011, p. 155) to meet the company's strategy (Koenig, 2012).

Thus, as mentioned above by Gartner's definition, whether the source of knowledge is the license, database, documents, or knowledge expert locator, these assets will go through this cycle in order to be created and utilized, and add value.

When we deal with KM, it is often technology related. It involves information technology and its infrastructure to facilitate the knowledge management system (KMS). Yet, it does not always have to be with information system. Technology will enable it and make it easier to execute the tasks and processes of KM. Other important dimensions in KM are culture and leadership. Culture has to establish an organization that is receptive to exchanging and sharing the knowledge. Leadership is important to push and establish the need for KM with the supporting culture and technology implementation. For KMS to be really effective, the culture of the organization needs to change from a knowledge-hoarding environment to a knowledge-sharing environment. This requires a drastic change in the way individuals are evaluated (Harrington, 2011).

The management process to manage the knowledge is as follows (Frost, 2010):

1. *Knowledge discovery and detection*: The knowledge that a firm possesses all over the organization, as well as the patterns in the information available that hide previously undetected pockets of knowledge.
2. *Knowledge organization and assessment*: To determine what resources they have at their disposal and to pin point strengths and weaknesses, management needs to organize the knowledge into something manageable. The knowledge organization involves activities that "classify, map, index, and categorize knowledge for navigation, storage, and retrieval" (Botha et al., 2008).
3. *Knowledge sharing*: Knowledge sharing can be described as either push or pull. The latter is when the knowledge worker actively seeks out knowledge sources (e.g., library search, seeking out an expert, collaborating with a coworker, etc.), while knowledge push is when knowledge is *pushed onto* the user (e.g., newsletters, unsolicited publications, etc.).
4. *Knowledge reuse*.
5. *Knowledge creation*: Continuous transfer, combination, and conversion of the different types of knowledge, as users practice, interact, and learn.
6. *Knowledge acquisition*: Refers to the knowledge that a firm can try to obtain from external sources.

KM Life Cycle

Installing a KMS is not easy; it requires that the total organization undergo a transformation that includes its culture, structure, and management style, but the results are well worth the effort. However, before we talk about the KMS, let us look at the KM life cycle. In its very simplest term, it is made up of six phases (see Figure 15.2).

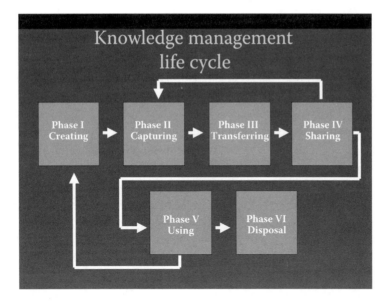

FIGURE 15.2
KM life cycle.

Capabilities are underpinned by knowledge. Therefore, organizations that seek to improve their capabilities need to identify and manage their knowledge assets.

Bernard Marr, Gianni Schiuma, and Andy Neely

"Intellectual capital—Defining key performance indicators for organizational knowledge assets."

***BPI Journal*, vol. 10, no. 5**

- Phase I—Creating knowledge
  Information and knowledge creation can and does occur at any place, which includes all parts of the organization and the outside world. It is the result of an individual's creative thoughts and actions. All individuals must be encouraged to become involved in contributing and increasing the availability of information and knowledge.
- Phase II—Capturing knowledge
  This is the act of preserving knowledge by identifying its value. This includes both hard (explicit) and soft (tacit) knowledge. This is so crucial that IBM hired outside journalists to interview their own people to capture how they made the decisions that led to successful outcomes.

Consequently human factors, the main driver of knowledge management success, has come second to information technology in many of the real world practices.

C. Carter and H. Scarbrough, "Toward a second generation of KM,"
***Education and Training*, vol. 43, no. 4/5, pp. 215–224**

- Phase III—Transferring knowledge
  This is the act of transforming knowledge inputs into a standard format that can be addressed by the stakeholders. It includes organizing the data into subject matter groups to meet the needs of the users. Furthermore, enabling processes must be put in place that will prompt the system to update the most current knowledge.

Intellectual workers enrich human knowledge both as creators and as researchers; they apply it as practitioners, they spread it as teachers, and they share it with others as experts or advisors. They produce judgments, reasoning, theories, findings, conclusions, advice, arguments for and against, and so on.

R. Cuvillier, *International Labor Review*, vol. 109, no. 4, pp. 291–317

- Phase IV—Sharing knowledge

Success depends on a clear strategic logic for knowledge sharing and it really depends on culture, that an organization should share and use knowledge automatically, and overcome the hoarding and trust issues.

J. Starr, 1999

The knowledge-sharing phase is the most important phase of the life cycle. People throughout the organization must be willing to share their knowledge and experiences if the KMS is going to succeed. However, there are many reasons why people do not want to share their knowledge. Here are some of them:

- "I am valuable because I know something that no one else knows."
- "I am not rewarded for sharing."
- "Sharing is a waste of time."
- "People should be able to think it out themselves."
- "I am too busy to share information."
- "It is not worth the time."

- "The timing is not right."
- "It takes too long to find out where to go to get the information."

The truth of the matter is that people are afraid of losing their personal competitive advantage if they share their knowledge.

In this phase, sharing knowledge refers mainly to the distribution of knowledge that has been already collected. It often takes place using a portal, but equally or more important, is using the social aspects of sharing knowledge, such as face-to-face meetings, virtual chat rooms, and building trust through personal communication methods.

Share your knowledge. It's one way to achieve immortality.

H. James Harrington

- Phase V—Using knowledge
  This is where the KMS pays off. Its big advantage is that the sharing of past experiences and knowledge helps prevent errors from occurring, and in creating new and better answers to the organization's opportunities. It is also during this phase that new knowledge is created and fed back to phase I.
- Phase VI—Disposal of knowledge
  As new ideas are created, better ways are developed, and best practices change, so it is important to keep the knowledge warehouse purged of any obsolete information and past best practices. This must be done with care, for sometimes historical data is very helpful as new approaches prove to be unsound.

Organizational capabilities are based on knowledge. Thus, knowledge is a resource that forms the foundation of the company's capabilities. Capabilities combine to become competencies and these are core competencies when they represent a domain in which the organization excels.

C.K. Prahalad and G. Hamel,
The Core Competence of the Corporation

Developing the KMS

The processes are

- *Using, enacting, executing, exploiting, etc.*
- *Communicating, deploying disseminating, sharing, etc.*

- *Compiling, formalizing, standardizing, explicating, etc.*
- *Appraising, evaluating, validating, verifying, etc.*
- *Acquiring, capturing, creating, discovering, etc.*
- *Evolving, improving, maintaining, refreshing, etc.*
- *Storing, securing, conserving, retaining, etc.*

A. Macintosh, author in *Knowledge Management,*
***Knowledge Organization, & Knowledge Workers*, 1998**

There is an overwhelming amount of knowledge within most organizations. To try to manage all of the organization's knowledge would be too time-consuming and costly. What most organizations do first, when they are considering installing a KMS, is to define what KM categories are most important to the organization and then concentrate the KMS on these key performance drivers. Often the KMS is designed to bring together the knowledge related to the organization's core capabilities and competencies.

The KMS depends on people using the technology to effectively share knowledge and experiences. It is very important to remember that people are the source of all knowledge, and technology is just an enabler that helps with the analysis and dissemination of the knowledge.

The following is a list of KMS-enabling modules, as Microsoft defines them. Basically, the information technology (IT) system helps handle information and transfers it into knowledge.

- Communities, teams, and experts
- Portals and search
- Content management (publish and metadata)
- Real-time collaboration
- Data analysis (data warehousing and business intelligence)

There are five major components required for a Workplace Neural Network (WNN) to enable measurable success. In our consumer-driven marketplace, it is essential that the required information is available, when it is needed, to satisfy the needs of the customer. The entire WNN system must be designed to meet each organization's unique needs. An organization that succeeds will provide an incentive system that provides multiple incentives for both sharing and seeking knowledge.

At a minimum, a WNN must provide communication with accurate information. It should provide the ability for individuals to acquire additional coaching or help from subject matter experts. In addition, this information, coaching, and support must be able to be captured and made available in near real time. Further enhancements should include the ability to track and test the validity of the knowledge, frequency of use, relevance to how work is done today, and other business objectives.

Knowledge that is no longer relevant or useful should be purged. The five major components of a KMS are

- Content (includes subject matter expert, governing body, data, information, knowledge in any form from any source, and vetting process)
- Availability (includes formatting, media, design, authoring, style, and usability measure)
- Collaboration management (includes interaction, storage, gap analysis, delivery tracking, and measurement)
- Generators (includes platform, systems, applications, etc.)
- Delivery style (includes location, method, pipes, hardware, and other software)

Content

Large consulting firms like Ernst and Young were early adopters of KMS as they hoped to capture the practical knowledge and experiences of their senior staff in order to leverage this knowledge throughout the less-experienced consultants. In doing this, they faced many obstacles. One of the biggest challenges was to define exactly what knowledge should be captured. Knowledge is useless unless it is relevant to the user's needs. It is often difficult to predict what knowledge is relevant and worth capturing unless you are able to anticipate the specific applications. Without taking time to accurately define how the knowledge will be applied, the KMS can quickly become an all-encompassing database, which is not useful.

The content of any organization's KMS should consist primarily of the data, information, and knowledge that provide the organization with its competitive advantage. The KMS should be reviewed on a regular basis by the organization's subject matter experts, as well as against any appropriate governing bodies requirements. The rating for all content within the WNN should be easily accessed for relevancy, currency, and other relevant details.

Everyone in the organization should be able to submit information for review and grading. This does not mean that everyone should be able to see any information they want, whenever they want. However, it does mean that anyone who has a legitimate business need should have access to the information when it is needed. Remember also that it is important to develop security levels for the KMS. In fact, biometrics should be used to provide more realistic security protection than on most networks today.

A KMS should include content from outside the organization, when appropriate. This often includes packaged training courses, white papers, and other pertinent data, information, or knowledge. It is the wisdom of the organization that makes this meaningful and how it is used in the organization is what provides the competitive edge.

Availability

Once the organization has defined what type of knowledge it needs to collect and disseminate, it then should define who the users of the KMS will be and how they will use the information. This is often a much more difficult task than one would estimate, at first glance. It is usually quite easy to define who will use the KMS, but it is much more difficult to determine how they will use the KMS. The *who* defines where and how you will communicate with the stakeholders. The *how* defines the makeup of the database and the content within each knowledge silo. Classification is how knowledge within the KMS is organized, searched on, and located. The classification taxonomy or schema is the most important part of the technology application. It defines how the knowledge is grouped and related.

As you can see, this presents a major challenge in getting agreement on how the output from the KMS will be presented and what information needs to be contained within the system. When the KMS is implemented, it will result in a whole new culture of information management throughout the organization. The goal is to have a knowledge-driven information network that will help solve organizational problems effectively, and to give the organization a significant competitive advantage.

Availability refers to how the knowledge is made available. It may be on an intranet, extranet, Internet, or other network. To be useful, the network must be easily available to all employees who need access to knowledge in their daily work. An effective KMS includes easy-to-use content authoring and publishing solutions. It should accept all needed formats such as

HTML, Flash, power points, streaming video, and so forth. All custom content should be developed to the Aviation Industry Computer Based Training Committee (AICC), Sharable Content Object Reference Model (SCORM) or other standards, so that it *plays nicely* with other systems. In addition, easy-to-control templates should provide a consistent look and feel. A robust KMS includes interactive simulations, case studies or games, online skills practice, testing, and feedback. Content publishing must be simple and require no programming or other special skills.

An effective KMS is a significant investment, but it will more than pay for itself as it increases your competitive advantage. To be effective, it must be able to track who has participated, what they have learned or taught, feedback on individual job performance, and calculate return on investment. Your KMS must be able to provide reports and tools that define a comprehensive and dynamic tracking system that evaluates how the KMS is being used and the value it delivers.

Collaboration Management

A KMS requires collaboration, interaction, online learning, and coaching on an as-needed basis. This includes small-group meetings, virtual training, working sessions, and knowledge-growing sessions, just to name a few. It must be usable, manageable, and scalable. The KMS provides an infrastructure for managing, tracking, sharing, archiving, and delivering strategic business knowledge across the extended organization. The data, information, and knowledge must contain appropriate links to organization-specific resources, including human resource and enterprise resource planning systems. Both the individual's knowledge gaps and organizational resource gaps need to be identified.

In addition to cataloging the organization's data, information, and knowledge, a robust KMS also tracks all access and its impact on how well the business needs are being met and on the return on investment. Using the KMS enables skill searches, employee assessments, applicant screening, and more.

Generators

A robust KMS will be able to be accessed from any appropriate platform and operating system. It will incorporate multiple forms of content, including PowerPoint, multimedia, web-based material, streaming video, shared whiteboard, and shared software applications. Furthermore, it will provide voice-over Internet protocol audio conferencing, live-application demonstrations, polling, real-time feedback and response, web touring, text chat and so forth. In addition, AICC and SCORM standards should be accommodated. It will also accept proprietary formats, as required.

Delivery Style

A KMS easily and conveniently delivers via appropriate network access. In addition, it must be able to be easily used to capture data, information, and knowledge from face-to-face encounters. A KMS must be able to be configured to work under your network circumstances from very narrow bandwidth to very high-speed lines. An organization must have the ability to communicate to its audience at the time and in the way that is most appropriate and continue the rich interaction and sharing among employees.

Tacit (Soft) Knowledge Flow

The most important and most difficult part of any KMS is the way soft knowledge flows through the organization and how it reaches its target audience. Although one of the key objectives of all KMS is to transform soft knowledge into hard knowledge (explicit), few KMS ever transform more than 20% of the soft knowledge into hard knowledge. One of the problems with soft knowledge is that it is often interpreted incorrectly by the individuals that it is being communicated to. The problem with verbal communications is that in the communicator's mind, communication is very specific and correct, but frequently the target audience misses the main point or misunderstands the communications. Verbal communications is the major way that soft knowledge is transferred from person to person. To have an effective soft KMS, you must understand the verbal communications process and the risks that are related to it. The biggest challenge that faces the KMS designers is how to error-proof the communication process.

To understand this better, we will review the 18 steps that a simple communication cycle consists of (see Figure 15.3).

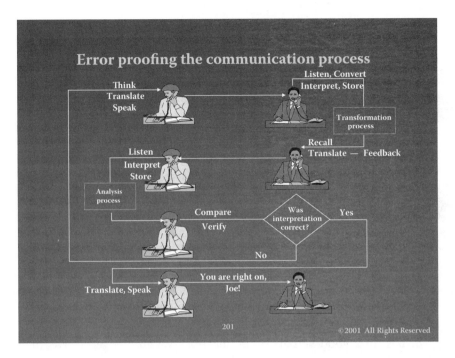

FIGURE 15.3
Verbal communication process.

Step 1. The verbal communication process starts with an individual having a thought that is greatly influenced by the culture, education, past experience, feelings, and environment that the individual is subjected to at the moment the thought was generated.

Step 2. This thought now has to be transformed into words.

Step 3. The individual finds a person that he or she wants to share the thought with and make that connection. In the example, the communication was made by way of a phone call. The fact that the communication is done by phone reduces the possibility that it will be transmitted correctly because the receiver cannot see the expression on the sender's face or his or her other body language. Communications over the phone are never as good as face-to-face communications, but the phone is usually much better than over the Internet, unless you need the communication documented.

Step 4. The communicator expresses the thought in a language that he or she believe the targeted individual (receiver) will understand.

Step 5. The sound waves travels through the communications medium, which downgrades the sound by reducing volume or distorting the sound quality.

Step 6. The *receiver* receives the sound waves and converts them into electrical pulses that travel through the nervous system to the brain where they are stored.

Step 7. The stored electronic pulses are then interpreted by the receiver based on his or her culture, education, past experience, feelings, and the environment that the individual is subjected to at the moment that the sound waves are received. This interpretation in the form of thoughts is then stored in the brain. (Note: The environment that the sender is in and the receiver is in can and often is very different. The sender's environment may allow him or her to be very focused but the receiver may be in the middle of doing something that distracts from his or her concentration, i.e., the kids may be acting up, the television may be playing, someone may be sitting at the receiver's desk and he or she may not want to keep them waiting.)

Step 8. In a good verbal communication system, the receiver is then called on to communicate back to the sender his or her understanding of the input. To do this, the receiver needs to transform the stored thought into words that he or she communicates back to the sender in a language that the sender will understand.

Step 9. The sound waves travel through the communication medium that degrades the sound by reducing volume or distorting the sound quality.

Step 10. The original sender receives the sound waves and converts them into electronic pulses that travel through the nervous system to the brain where they are stored.

Step 11. The stored electronic pulses are then interpreted by the original sender based on his or her culture, education, past experience, feelings, and environment, and is then stored back in the brain in the form of a thought.

Step 12. The sender then needs to compare the original thought to the ones that he or she received from the receiver to determine if the receiver interpreted the original thoughts correctly.

Step 13. If the thoughts were interpreted incorrectly, the process reverts back to step 1 and starts over. If the thoughts were analyzed as being interpreted correctly, the sender must translate the concluding thought into words.

Step 14. The original sender expresses the thought in a language that he or she believes that the target individual will understand.

Step 15. The sound travels through the communication medium that degrades the sound by reducing volume or distorts the sound quality.

Step 16. The receiver receives the sound waves and converts them into electronic pulses that travel through the nervous system to the brain where they are stored.

Step 17. The stored electronic pulses are then interpreted by the receiver based on his or her culture, education, past experience, feeling, and environment, and they are then stored in the brain.

Step 18. The receiver takes appropriate action based on his interpretation of the feedback.

One of the things that make effective verbal communication so difficult is that it is more than just words being transmitted to individuals or groups of individuals; it includes also the tone of voice, the inflection on the words, and the hesitation on the delivery. All of this adds up to interpreting and obtaining the best knowledge. To add to the complexity of verbal communication, much of it occurs face-to-face. In these conditions, additional complexity is added as the nonverbal communication often tells more about the content of the communication than the words themselves communicate. Nonverbal communications is another completely separate communication system from the one that we have just discussed. It is one that we all need to understand and use. All KMS must include training in how to communicate to effectively transmit knowledge from one source to another. Part of this training has to be nonverbal communication techniques. We like to think that communications with the technologies that are available today is getting better. The truth of the matter is, there is a lot more of it but the quality of the communication is much poorer than when we used to meet face-to-face. Today, we have quantity but decreased quality and understanding.

Explicit (Hard) Knowledge Flow

We are drowning in information but starved for knowledge.

John Naisbitt, author, *Megatrends*

The KMS (Figure 15.4) is a filtering and analysis system. The raw materials that go into this system are data from the many measurement points

Knowledge management system

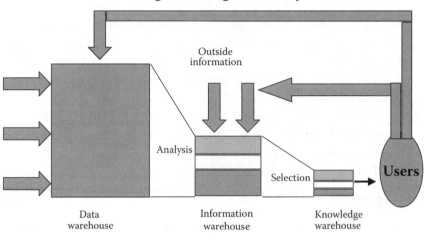

FIGURE 15.4
Knowledge management system.

within and outside the organization. The data is accumulated and stored in a data warehouse. The data in the data warehouse is then analyzed and distilled into information that takes the form of reports, statements, bills, etc. This information is typically stored in an information warehouse. The information warehouse also receives information from many other sources, like minutes of meetings, external reports, books, literature reviews, conferences, and papers, and from tacit knowledge that is converted into explicit information. Do not underestimate the amount of excellent information that can be collected from external sources (outside information).

> Books are not lumps of lifeless paper but minds alive on the shelves. From each of them goes out its own voice ... and just as the touch of a button on your stereo set will fill the room with music, so by taking down one of these volumes and opening it, one can call into range the voice of a man far distant in time and space, and hear him speaking to us, mind to mind, heart to heart.
>
> **Gilbert Highet, literary critic**

As Figure 15.4 indicates, the amount of items in the information warehouse is far less than is contained in the data warehouse, even though information, which is not based on the data in the data warehouse, is included because it is distilled data. The information warehouse is then organized into information silos that contain the information related to the specific knowledge subjects. This is usually done electronically using intelligent agents or network mining.

> The need to acquire knowledge from outside—which you may call benchmarking—and the need to acquire knowledge from inside—the sharing of best practices—and then make it portable is a universal concept that will remain long after modern metaphor has changed.
>
> **Steve Kerr, chief executive officer, GE**

The five Cs that differentiate data from information are condensation, calculation, contextualization, correction, and categorization.

- Condensation—Data is summarized in a more concise form.
- Calculation—Analysis of data similar to condensation of data.
- Contextualization—You know why the data was collected.
- Correction—Errors have been removed.
- Categorization—The unit of analysis is known.

The information within each silo is then reviewed to determine if it represents new or added-value information that should be added to the knowledge warehouse. This is often accomplished by subject matter experts. These same subject matter experts will maintain the knowledge warehouse, updating it as new information becomes available, and removing obsolete records. For example, the best practices for an individual knowledge subject are continuously changing so that the old best practices need to be removed and replaced by the current best practices. The knowledge warehouse never has raw data stored in it.

The output from the knowledge warehouse (repository) usually takes the form of a restricted portal that is set up for the users of the knowledge category. The knowledge warehouse can be considered the *bank* of our intellectual capital (see Figure 15.5). The knowledge warehouse usually contains structured electronic repositories of knowledge, either structured document-based knowledge, information discussion-type knowledge, or repositories of *who knows what*.

Explicit knowledge often does not communicate the true meaning of the experience. Without rich cultural and conversational content, documents lose much of their real value. The inflection in the voice, the

Knowledge database

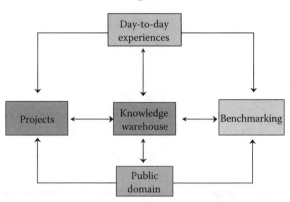

FIGURE 15.5
Knowledge database.

expression on the face, the twinkle in the eyes, and the body movement are all part of the total knowledge that is being communicated. Face-to-face communications is always better than documents and reports. It is even better than phone calls or e-mail. It is for this very reason that the knowledge community is using video conferencing so much. The problem with video conferencing is it normally produces only tacit (soft) knowledge, rather than explicit (hard) knowledge. To help with this problem, we ran across a software product, Visual Communicator Studio by Serious Magic (www.seriousmagic.com). It is a simple tool that captures and shares multimedia presentations in a way that enriches personal knowledge. All you need to use this tool is one or two video cameras, a microphone, and a laptop to document the meeting. It combines PowerPoint presentations, whiteboard, and screenshot images. When you select a PowerPoint presentation, you can hear the conversation that was related to the slide, as well as see any notes that were recorded by the participants.

KM is a tool that provides input of knowledge asset to the innovation process in all stages for individuals who will need the right knowledge assets to achieve the business strategy objective in order to stay sustainable. KM is dependent on technology and IT specifically, yet also has its own dimensions and processes to make it adaptable and useful when disseminating knowledge that leads to innovation that is systematic and continuous.

EXAMPLES

AIR PRODUCTS USES CHEMREG
TO MEASURE COMPLIANCE

Profile: With more than 2000 active products as well as several hundred new products each year, Air Products was very concerned about measuring its compliance to the U.S. Department of Transportation (DOT) requirements surrounding the shipment and transportation of hazardous materials. The company had a manual process of generating shipping descriptions for bills of lading, labels, and materials safety data sheets, which was very labor intensive, time consuming, and error prone. Their business partners were getting less and less tolerant of 10–30 days required to establish a new product, and constant scrutiny required an environment of continuous improvement in safety, health, and environmental performance.

KM Strategy: In 1993, the company created a special KM task team to create a system for measuring compliance to new regulations, HM-181. This was the birth of CHEMREG, an internally developed software system having two separate but *integrated* applications:

- Rational product database containing physical characteristics on all of the commercial products and many experimental ones in the Chem Group, where data is inputted by various business areas as well as laboratory personnel.
- Knowledge-based system that generates information, such as shipping descriptions and instructions, based on product data and regulatory rules, in much the same way that an expert system would.

Success Story: All appropriate regulatory information has been incorporated into the knowledge-based portion of CHEMREG, allowing it to produce accurate shipping descriptions for the following agencies: DOT, International Air Transport Association (IATA), and International Maritime Organization (IMO).

The CHEMREG team developed a new process for printing planning document instructions that assist the shipping locations in areas of proper labeling, placarding, and packaging instructions. A new global information system was added in 1994 to gather data more quickly and inform individuals of new shipping descriptions for newly established products. This reduced the cycle time from seven to 30 days down to two days for the entire process. The measurement system also has a query capability that allows users to perform *what-if* analysis on product data and regulatory compliance.

Air Products has recently modified its knowledge-based expert system to accommodate another cycle of regulatory changes by the DOT, IMO, and IATA. This was followed by the automatic generation of Material Safety Data Sheets (MSDS). Using product databases for product characteristics, an expert system determines appropriate phraseology and content for the MSDS. Eventually, CHEMREG is expected to become the focal point for distribution and environmental-, health-, and safety-related regulatory issues around the world. Future topics to be addressed by the CHEMREG system include generating the content of the MSDS based on European regulations, including ISO 14000 and related quality system standards.

Source: Harrington, H.J. and Voehl, F. *Knowledge Management Excellence.* Chico, CA: Paton Press, 2007. Strategy Associates Inc., with permission.

REWRITING THE UNWRITTEN RULES AT RUTGERS UNIVERSITY

Background: Rutgers University (State University of New Jersey) has 50,000 students and 5000 researchers and faculty scattered among a dozen campuses in the state.

KM Strategy and Structure: To facilitate communication, Rutgers is spending $100 million to lay high-speed fiber optics that will link all of its classrooms, offices, laboratories, and dormitories. Professor Wise Young, director of the Neuroscience Center at Rutgers, is currently engaged in establishing a new collaborative research facility at the university. The center is being designed for knowledge sharing, not only for laboratories within Rutgers but in conjunction with more than 60 other research laboratories around the world. Every part of the laboratory is specialized for efficient communication and sharing of visual, audio, and numerical data, as well as real-time personal interactions. Even many of the laboratory instruments are designed to allow groups of users to access them remotely.

The center will use primarily two types of software for remote collaboration: Timbuktu Remote and CU-SeeMe. Timbuktu allows people to work on computers as if they were physically sitting in the central laboratory. CU-SeeMe allows up to eight individuals to video conference with each other. "Both programs are cheap and powerful, and are flexible enough to accommodate a variety of Internet bandwidths," Young said.

Success Story: Young's future purchase plans include a microscope (the Zeiss 510) with state-of-the-art video cameras that can collect and store slide images in the form of movies. "Instead of throwing a slide on the wall to show an image, a lecturer can access a server over the Internet to play a high-resolution video movie," Young said. "In Osaka (Japan), for example, there is a high-voltage electron microscope that can be used over the Internet."

Hurdles: Costing millions, such microscopes and facilities are beyond the means of individual scientists. It is also important to allow such a facility to be used around the clock, to make the most of the investment. "This is the wave of the future," Young said. Such cutting-edge technological advancements require people to work together in new ways, as well. "One has to learn to collaborate. It doesn't come naturally," he said.

Outlook: "Because our academic and business organizations are predicated on competition, there is a natural reluctance to share and to give. But it can and does happen. As a member of a team, you begin to realize that the whole is greater than the sum of the parts. You realize that being part of the team allows you to do things that you would never otherwise think of."

SOFTWARE

KMS is not software in itself. The system has enhanced results when combined with the right software.

- Content Management System (Sharepoint)
- Salesforce

REFERENCES

Frost, A. Knowledge management processes. Retrieved from KMT, an educational KM site: http://www.knowledge-management-tools.net/knowledge-management-processes.html, 2010.

Harrington, H.J. *Knowledge Management Excellence: The Art of Excelling in Knowledge Management.* Chico, CA: Paton Professional, 2011.

Koenig, M.E. What is KM? Knowledge management explained. Retrieved from KMWorld: http://www.kmworld.com/Articles/Editorial/What-Is-.../What-is-KM-Knowledge-Management-Explained-82405.aspx, 2012.

McAdam, R. Knowledge management as a catalyst for innovation within organizations: A qualitative study. *Knowledge and Process Management*, vol. 74, pp. 233–241, 2000.

Nonaka, I. and Toyoma, R. A Firm as a Dialectical Being: Towards a Dynamic Theory of a Firm. *Industrial and Corporate Change*, vol. 11, pp. 995–1009, 2002.

16

Market Research and Surveys for Innovations

Fiona Maria Schweitzer and Julian Bauer

CONTENTS

He who asks is a fool for five minutes. He who does not ask is a fool forever.

Chinese proverb

DEFINITION

Marketing research can be defined as the systematic and objective identification, collection, analysis, dissemination, and use of information that is undertaken to improve decision making related to products and services that are provided to external customers.

You are never successful until the market says you are.

USER

This tool can be used by individuals but its best use is with a group of four to eight people. Cross-functional teams usually yield the best results from this activity.

OFTEN USED IN THE FOLLOWING PHASES OF THE INNOVATIVE PROCESS

The following are the seven phases of the innovative cycle. An X after the phase name indicates that the tool/methodology is used during that specific phase.

- Creation phase X
- Value proposition phase X
- Resourcing phase
- Documentation phase
- Production phase
- Sales/delivery phase X
- Performance analysis phase

TOOL ACTIVITY BY PHASE

- Creation phase—Market studies are usually used during this phase to define customer needs and expectations and also to define the size of the market.
- Value proposition phase—Market studies are used here to validate the assumptions related to customer requirements and value-added considerations. Size of the market, customer expectations, and price point are major considerations in preparing the value proposition. The best way to acquire this type of information is through a statistically sound market study.
- Sales/delivery phase—Market studies are frequently used to define the best sales and marketing strategy.

HOW TO USE THE TOOL

Marketing research can be defined as the systematic and objective identification, collection, analysis, dissemination, and use of information that is undertaken to improve decision making related to identifying and solving problems in marketing. In other words, the main purpose of market research is to collect and provide valuable information for a company for a better decision making (Malhorta, 2009). But how is market research connected to innovation management? To answer that question, it is important to understand the nature of innovation. Innovation is about doing something new and unknown. It contains opportunities and possibilities, as well as risks and uncertainties. In general, innovation projects are faced with three main dimensions of uncertainty: resource/organizational uncertainty, technology

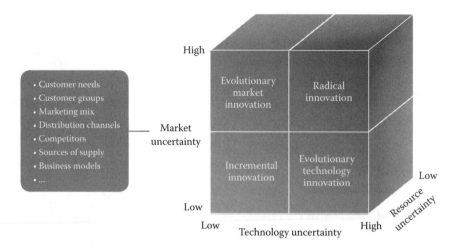

FIGURE 16.1

Technology and market uncertainty. (Based on Gaubinger, K., Rabl, M., Swan, S., and Werani, T. *Innovation and Product Management—A Practical Approach to Reduce Uncertainty*. Wiesbaden, Germany: Gabler/GWV Fachverlage GmbH, 2014.)

uncertainty, and market uncertainty (see Figure 16.1). Market uncertainty is related to all risks regarding customers, market, competition, or the macro-environment. Market research aims to reduce market uncertainty.

For that reason, market research plays an important role in innovation management (Gaubinger et al., 2014). It offers important information about the customers' latent needs and enables the company to read the customers' mind. This knowledge should be used to reduce market uncertainty and make the right decisions in new product development.

The main task of innovation management in a company is to organize new product development in an effective and efficient way. For this end, innovation management performs strategic and operative tasks. Market research provides information to improve decision making regarding both tasks.

Market Research for Operative Tasks

The key operative task in innovation management is to guarantee an efficient innovation process, from idea development to concept and product development and finally to launch. A central tool is the *stage gate process*, which allows for a structured and systematic procedure of moving the idea through to the launch. The stage gate process is divided up into stages in which information from the two core areas of innovation management, research and development and marketing, is collected. At the following gates, this information is evaluated by an interdisciplinary team. The aim of this process is to make decisions based on the gained information to determine if the innovation project should be continued or not. This daily business should be done in an efficient way with regard to time, resources, and quality (Cooper, 2011). Market research plays a central role in each of these phases. The beginning of the process—the front-end stage—starts with tasks like trend scoping, idea generation, idea evaluation, concept generation, and concept evaluation. The middle—central stage—contains the phases for development, evaluation, and testing. The final launch is the back end of the process. In each of these phases, different methods of marketing research can be applied to reduce market uncertainty. The trend-scoping phase can be supported by methods like Delphi, empathic design, or future workshops. In the next phase—the idea generation phase—the *lead user* method could be used. The idea generation phase, for instance, can be contributed to by market research of idea competitions. In the idea evaluation phase, early customer feedback can be used to review the projects. The preferred ideas are developed in the concept generation phase in a more mature form accompanied by in-depth analysis. After that, the developed projects are reviewed, for instance, by focus groups in the concept evaluation phase. In development stage, a more detailed marketing concept is acquired. Various tests with samples, prototypes, and potential customer are necessary to approve the concepts. Before the final product launch is carried out, the retesting, validation, market tests, and testing on the marketing mix have to be done to secure a successful market entry (Gassmann et al., 2014).

Market Research for Strategic Tasks

Central strategic tasks in innovation management include defining the innovation strategy, managing the innovation portfolio, and planning unique and differentiated market offerings. The goal is to increase innovation effectiveness, understood as *doing the right things*. Hence, the aim is to choose the right innovation projects by exploring, planning, and steering. Market research for strategic innovation management can provide vital information to facilitate these tasks. For example, it can help identify weak signals in markets that provide new growth opportunities or it can detect changes in customer needs that give essential background information on the selection of the right innovation projects. Furthermore, market research provides data on the competitive situation, and it is needed to calculate the market potential for new innovation projects (Tidd et al., 2001).

The first step in carrying out a market research project—be it for operative or strategic innovation tasks—is defining which type of information is needed. On the basis of the information requirements, a decision has to be made on whether secondary data is sufficient or a primary study is required. Furthermore, it is necessary to choose either a quantitative or a qualitative research approach. Next, you have to choose the right instruments to gather the required information. A central instrument in primary research is a survey. For this reason, basic principles on writing survey questions are briefly discussed.

Using Primary Research or Desk Research

Primary research is done by collecting data for the first time for a special problem or purpose of the certain study such as survey data. Some typical applications for the innovation management in the early phases are indepth interviews or the lead user method. For the later launch phase, the example of the store or market test could be given.

The big advantage of this method is the exclusivity and best matching of the data. By developing the research design, the aim of the collection is precisely defined and the research problem is exactly addressed. A disadvantage is the high effort to collect the data. This might be time consuming and expensive. Therefore, researchers should always consider the use of secondary data. It might be that the present problem is not unique and someone has already gained some relevant knowledge.

In contrast to primary research, desk research is using secondary data that can be collected on the Internet, from trade organizations or government agencies. In the case of innovation management, one example would be a competitive analysis on the Internet. It represents a descriptive, market characteristic study form.

Another use of secondary data would be for netnography. This method is an ethnographic research (observation of the behavior of a certain group of people) on the Internet. It uses online communication flow to get information about the needs and desires of the participants. The main advantage of using secondary data is time and cost efficiency. For that reason, desk research should always be the start in every market research to identify the problem and get a better understanding for it. But secondary data can also be limited by its degree of fit, time accuracy, and compatibility (Malhorta, 2009).

Qualitative or Quantitative Market Research

The main purpose of qualitative research is to give insights and a deeper understanding of the research problem. Qualitative research is used when facing a situation of uncertainty. Qualitative research is based on a small sample size, and it is usually analyzed in a nonstatistical way. The aim is to get a deeper and richer understanding of motivations and reason for a certain problem.

For innovation management, two highly relevant methods can be mentioned. The first example is the focus-group interview. The purpose of this approach is to get information of the researchers' interest by listening to a group discussion of 8–12 people of a certain target market. In the free-flowing group conversation that is headed by a moderator, rich information findings can be achieved. As another example, in-depth interviews can be given. These loosely structured one-on-one conversations are unstructured and attempt to uncover motives, feelings, or attitudes toward a certain topic. One aim could be to get deeper insights of customer needs (Malhorta, 2009).

In contrast to qualitative research, quantitative research is seeking to quantify data in the form of numbers and figures. Typical for that method are large, representative samples and statistical analysis. The aim of qualitative research is to get generalized results from the sample to the population of interest. One application in innovation management would be a product test or prototype test via a quantitative questionnaire. The questionnaire consists of structured questions that are designed to quantify the consumer's opinion about a product. In the end, the different prototypes can be based on the ranks of consumer preference.

Additionally, *conjoint analysis* is a quantitative market research tool. It explores consumer preferences for different hypothetical product concepts. Each concept is composed of certain attribute levels. Through this artificial construction of concepts, different preferences can be directly linked to different attribute levels so that this method allows you to calculate exactly what consumers are willing to pay for different features and attributes (Schweitzer, 2014).

Frankly, the qualitative and the quantitative method both have pros and cons (see Table 16.1). On the one hand, qualitative research has the advantages of depth of the data, the evocation of creativity and intuition, and the exploration of latent attitudes and needs. Contrary to that, qualitative research does not cover a representative sample, is not quantifiable, and is

TABLE 16.1

Advantages and Disadvantages of Qualitative and Quantitative Research

Research Method	Advantages	Disadvantages
Qualitative research	+ Explores problems and motivations	
	+ Stimulates creativity	
	+ Generates a deep understanding for latent needs	− Nonrepresentative data
		− No statistical analysis possible
	+ Flexible during the collection phase	− Unstructured data collection
	+ Small sample size	
	+ Generalized results	
Quantitative research	+ Statistical analysis possible	− Large representative samples needed
	+ Structured data collection	− Inflexible during the collection phase
	+ Quantifies opinions	
	+ Base for decision making	

demanding a high level of data collection. On the other hand, quantitative research offers a broad and secure database, as well as representative and quantifiable results. In opposition to qualitative research, quantitative research is more structured and inflexible during the collection phase. In practice, both research methods are used very often in combination to get the best results. Qualitative research seeks for a richer understanding, whereas quantitative research tries to confirm a final decision based on figures. For those reasons, market research starts often with a qualitative research. After a basic understanding is achieved, the quantitative methods are applied (Malhorta, 2009).

Guidelines for Writing Good Surveys

Surveys are used in a variety of contexts in market research. They can be conducted via e-mail, telephone interviews, formal structured personal interviews, and self-administered studies. Irrespective of the form used, the ways in which the surveys are structured are very broadly similar. The aim of a survey is to collect data to answer the research question best. Therefore, the following guideline should be used to ensure the best results:

- First, the structure should include three major parts: the introduction, the body, and the basic data. The introduction should outline the general purpose and aim of the study. Furthermore, in this part, some questions can be used to check if the potential participant really belongs in the sample. In the body, the main questions are asked to solve the research question. For the last part, the basic data should include demographic data like age, nature, or income.
- Secondly, some general rules should be followed to ensure an easy understanding of the survey:
  - Use a clear and simple language style
  - Keep the questions short and to the point
  - Ambiguity and vagueness must be avoided
  - The questions should be neutral (not loaded, negative, or hypothetical)
  - Try to avoid assumptions
  - Do not overtax the participants
- Thirdly, different types of response format can be used:
  - Closed-ended questions—followed by a structured response
  - Dichotomous question—suggests two answers like *yes* or *no*
  - Open-end questions—do not suggest an answer; people can write what they wish
  - Unstructured questions—allow the participants to reply in a format they prefer
  - Measurement of the data—nominal, ordinal, interval, ratio
- Last, but not least, the whole survey should be pretested and potential problems eliminated to ensure a clear understanding and fluent responding (Proctor, 2005; Malhorta, 2009).

Overview: Market Research Methods—Innovation Process

Market research can support the innovation process to avoid market uncertainty in various ways. Qualitative and quantitative research, as well as primary and secondary data, can provide meaningful information to assist decision making in the different phases of an innovation process. Yet, not all of these methods qualify to the same extent for every phase of the innovation process. While qualitative methods are often useful in the early phases, quantitative methods are more valuable in later phases (Gassmann and Schweitzer, 2014). Figure 16.2 provides examples

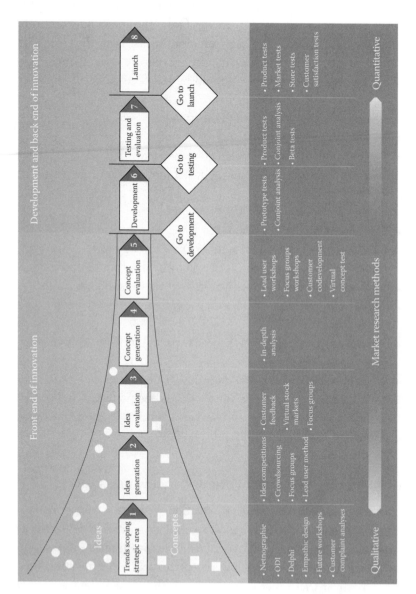

FIGURE 16.2
Examples of market research.

of market research methods that are typically used in the different phases (Gassmann and Schweitzer, 2014).

Market Research for Idea Generation and Evaluation

During the very first phase of the innovation management process, the main aim is to create ideas and then to confirm and evaluate these ideas by the customers to secure the market fit. For innovation management especially, latent customer needs are the most interesting. Latent customer needs are strong needs that cannot be satisfied by existing products or services. To create new innovative solutions, this information is most valuable.

Opportunity Screening and the Detecting of Latent Needs

Starting with opportunity screening, the very first step is the analysis of the company environment. In its macro-environment, the company is facing threats, as well as opportunities. These can be categorized into five major forces: political/legal, economic, ecological/physical, social/cultural, and demographic and technological. Market research can support the detection of these opportunities by providing qualitative and quantitative information on these five environments. For example, market research can gather trend data on consumer lifestyles, habits, values, and beliefs to register potential sociocultural change.

The microenvironment of a company is defined by another six factors: the company, suppliers, marketing channel firms, customer markets, competitors, and the public. The relevant factors for opportunity screening are, in that case, the customer, the market, and the competitors. A relevant qualitative market research method is the estimating of future demand and buyers' intentions via sales force or experts opinions (Kotler et al., 2009).

All these methods of screening of the company's environment should give a better understanding of the customer needs. But there are also more direct ways to get customer needs. Two of them would be the generation via idea competitions or crowdsourcing. Idea competitions stimulate participants' ambition and creativity. Different forms for realization like the media environment (online or offline), problem specification, degree of elaboration, target group, duration, and evaluation are possible (Schweitzer, 2013).

Another method is the screening of customer complaint data. This form of input illustrates a key source for new ideas. It offers the company a way to get insights into the world of the customer, their desires, needs, and requirements. About 35% of all innovative solutions are based on customer complaints (Vahs and Brem, 2013).

Market Potential Evaluation through Secondary Research, Expert Interviews, and In-Depth Interviews of Customers

Once the first concrete ideas have been generated, the potential of these has to be evaluated in the next phase. One major key indicator of the future success of the new product is the market potential. Two qualitative methods are often used to gain knowledge about the prospective market situation.

The first method is carried out through expert interviews with dealers, distributors, suppliers, marketing consultants, and trade associations. Expert interviews can be organized in a group discussion or in individual sessions (Malhorta, 2009).

A special form of gaining knowledge from experts is the Delphi method. In this method, data is gathered through individual expert interview and then the results are shown in an anonymous form to all experts who are asked to reevaluate their ideas and assessments. The Delphi method is performed in an iterative process (at least three rounds) with feedback rounds to find one consensual outcome in the end (Schweitzer et al., 2014).

Idea Evaluation in Focus Groups

Besides the market potential, other aspects might be necessary to forecast the success of new product ideas. Focus groups in which the raw concept for the new product is discussed can indicate the likely success or failure of an idea. Eight to twelve people led by one or two moderators evaluate the ideas with different techniques. Furthermore, focus groups can give the innovating company ideas for new product solutions (Proctor, 2005).

Market Research for Concept and Product Development and Testing

In the next phase, the selected ideas are elaborated to advanced concepts, samples, prototypes, and first serial products, and again evaluated by potential customers. The aim in this phase is that customers support the product development (customer co-development), and the concept is evolved to series-production readiness. The test persons can vary from average users up to well-experienced lead users. Qualitative and quantitative methods are used for data collection. Early concepts are usually tested with qualitative methods to receive first information. More evolved concepts and prototypes ate often tested in a quantitative way.

Traditional Concept, Prototype, and Product Testing

The concept evaluation phase is the last step of the front end of innovation. The different concepts and variants are tested and assessed. A concept is a further stage of an idea and contains a clear and precise description, the essential characteristics like the design, or the customer benefit of future products. The form of the concept presentation can differ. Product models, functional models, pictorial representations, or written or verbal explanations are possible to get a high benefit–cost ratio. The main purpose of the concept test is to get an idea of the customers' reactions, feedbacks, or buying intentions. In a virtual concepts test, customers can express their preference concerning product and price variants. This method has a considerable cost and time advantage over classical physical prototyping (Schweitzer et al., 2013).

To support the development phase, testing with physical prototypes are done in iterative loops. In the software industry, this approach is very common. With iterative prototyping, it is easier to achieve the product goal in a much shorter time frame (Cooper, 2012).

Another efficient method for testing prototypes and developing products is a user or product clinic. In this clinic, the subjective perceptions of prototypes or already existing products by customers is analyzed. This method is focusing on the product characteristics. The aim is to get information about customer satisfaction, potential market segments, and the acceptance of the product. A well-known example out of the automotive industry is a car clinic, where newly developed cars are tested (Vahs and Brem, 2013).

Lead User Workshops

Lead users are ahead of trends and future market needs, and can gain great benefit for new product development. Additionally, lead users are highly motivated to solve these special needs for their own purpose. Those certain characteristics make them important contributors to new product development. At the same time, they make it difficult to identify this special type of users. Methods like screening, pyramiding, netnography, or idea competitions could be used to select them. Lead users can be implemented into product development in different ways like interviews or workshops. Because of their special expertise, they can be a source for customers' needs as well as for product and solution information (von Hippel, 1986; Reichwald and Piller, 2009).

Toolkit Method

Toolkits for user innovations are user-friendly design tools that enable customers to develop their own product solutions. The customers are following an iterative trial-and-error workflow in a certain solution space. By this method, the design and innovation tasks are shifted to the customer himself. At the same time, the technical feasibility is guaranteed by the toolkits, and the market orientation is ensured by the input of the customer. Toolkits can be used in focus groups to find ways of improving product concepts and to learn which attributes of a product contribute value (von Hippel and Katz, 2002).

Key Questions—Concept Tests

Before performing a concept test, it is important to measure product dimensions and ask the participants questions about what they like and dislike about the product concept, whether it satisfies an important need, and how big this need is. This information is important to change certain features, to make the product fit the needs better, and to get early information on product acceptance. Furthermore, information on the perceived uniqueness compared with competitive products should be in the test questionnaires, as this again says something about the likely interest in the new product. Moreover, information on the potential target group has to be gathered, to be able to say consumers from which sociodemographic background and with what lifestyle are more likely to buy the product than others. Background information on the respondents together with purchase likelihood information helps marketers and innovation managers to make first forecasts of potential demand of the final product.

Generally speaking, the following key questions should be answered in a concept test:

1. *Communicability and believability*: How clear and believable are the benefits of the concept? Low believability and low clarity values indicate that the concept must be revised.
2. *Need level and perceived value*: How good is the concept at solving a specific problem or filling a need? How valuable is the potential end product for the customer? How strong is that need or how relevant is the problem? Basically more important needs or problems will lead to higher expected customer interest.
3. *Gap level*: Which other products solve this problem or meet this need? Fewer competitive solutions will bring about higher customer interest.
4. *Purchase intention and willingness to pay*: Do the respondents state that they would buy the product? Which price are they willing to pay? Information on purchase intention and price acceptance helps calculate potential demand and sales volumes.
5. *User targets, purchase occasions, purchasing frequency*: What gender, lifestyle, purchase history, etc., do those respondents who are interested in the product have? This information helps specify the target group (Kotler et al., 2009).

Prelaunch Market Research

Before the developed products are launched, they are normally analyzed in test markets. The aim is to get the very last precious information for the last phase. The gained information should help optimize and confirm the planned program to enter the market.

Market Testing and Test Markets

Market testing and test markets are controlled field experiments in which the product or the whole marketing program is evaluated in selected stores for several months. The purpose of these methods is to determine the market acceptance, on the one hand, and to test the variables of the market mix program, on the other hand. During the testing, researchers can vary these factors like promotional investments, pricing, product modifications, or distribution to define the best setup for the introduction. Besides the fine tuning of the marketing mix to use the resources in the most effective way, information concerning sales, profits, or potential product problems can be collected.

For the view of innovation management, these methods are the very last check before the new product is launched on the target market. The gathered information can be used for last-minute changes as well as for next product generations or product variants in the future (Malhorta, 2009; Cooper, 2011).

Quantitative Calculation of Market Potential

The market potential is the entire size of the market for a product during a specified period. In other words, the maximum possible market demand of a certain time period is the market potential (see Figure 16.3).

To calculate the market demand, some other key figures are used, like the number of buyers in the market (n), the quantity purchased by an average buyer per year (q), and the price of an average unit (p).

$$Q = n \times q \times p$$

By multiplying these factors, it is possible to quantify the market potential. For making forecasts about the future market situation, estimations about these figures are normally done by the sales force or branch experts (Kotler et al., 2009).

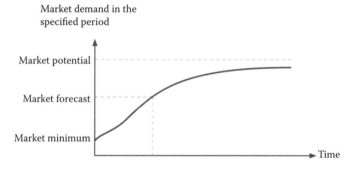

FIGURE 16.3
Maximum market demand.

EXAMPLES

See examples included in the section entitled How to Use the Tool.

SOFTWARE

No specific software recommended although there are a number of software packages commercially available.

REFERENCES

Cooper, R.G. *Winning at New Products: Creating Value Through Innovation.* 4th Ed. New York: Basic Books, 2011.

Gassmann, O. and Schweitzer, F. *Management of the Fuzzy Front End of Innovation.* New York: Springer Verlag, 2014.

Gaubinger, K., Rabl, M., Swan, S., and Werani, T. *Innovation and Product Management—A Practical Approach to Reduce Uncertainty.* Wiesbaden, Germany: Gabler/GWV Fachverlage GmbH, 2014.

Kotler, P., Keller, K.L., Brady, M., Goodman, M., and Hansen, T. *Marketing Management.* Essex, UK: Pearson Education Limited, 2009.

Malhorta, N.K. *Basic Marketing Research—A Decision-Making Approach.* Upper Saddle River, NJ: Pearson Education, 2009.

Proctor T. *Essentials of Marketing Research.* Essex, UK: Pearson Education Limited, 2005.

Reichwald, R. and Piller, F. *Interaktive Wertschöpfung.* Wiesbaden, Germany: Gabler Fachverlag, 2009.

Schweitzer, F. Integrating customers at the front end of innovation in management of the fuzzy front end of innovation. In: O. Gassmann and F. Schweitzer, eds. *Management of the Fuzzy Front End of Innovation.* pp. 31–48. New York: Springer, 2013.

Schweitzer, F. Chapter 18, Market research in the process of new product development. In: G. Praveen and B.E. Trusko, eds. *Global Innovation Science Handbook,* pp. 283–302. New York: McGraw-Hill, 2014.

Tidd, J., Bessant, J., and Pavitt, K. *Managing Innovation.* 2nd Ed. Chichester, UK: John Wiley & Sons Ltd., 2001.

Vahs, D. and Brem, A. *Innovationsmanagement: Von der Idee zur erfolgreichen Vermarktung.* Stuttgart, Germany: Schäffer-Poeschel, 2013.

von Hippel, E. Lead users: A source of novel product concepts, *Management Science,* vol. 32, no. 7, pp. 791–805, 1986.

von Hippel, E. and Katz, R. Shifting innovation to users via toolkits, *Management Science,* vol. 48, no. 7, pp. 821–833, 2002.

SUGGESTED ADDITIONAL READING

Kotler, P., Wong, V., Saunders, J., and Armstrong, G. *Principles of Marketing*. Essex, UK: Pearson Education Limited, 2008.

17

Organizational Change Management (OCM)

H. James Harrington

CONTENTS

You can manage change or it will manage you. The choice is yours.

H. James Harrington

DEFINITION

Organizational change management (OCM) is a comprehensive set of structured procedures for the decision making, planning, executing, and evaluation of activities. It is designed to minimize the resistance and cycle time to implementing a change.

USER

This tool is usually used by a project management team when working on a complex or important project.

OFTEN USED IN THE FOLLOWING PHASES
OF THE INNOVATIVE PROCESS

The following are the seven phases of the innovative cycle. An X after the phase name indicates that the tool/methodology is used during that specific phase.

- Creation phase
- Value proposition phase X
- Resourcing phase X
- Documentation phase X
- Production phase X
- Sales/delivery phase
- Performance analysis phase

TOOL ACTIVITY BY PHASE

- Value proposition and resourcing phases—During these phases, the plan for organizational change activities plays a key role in estimating project implementation time, impact, and risk.
- Documentation phase—During this phase, OCM activities are directed at communicating a vision related to the change and the impact that the change will have on the individuals who will be using the changed process.
- Production phase—The primary impact of OCM will be recognized during the production implementation phase of the proposed project and during the months following implementation where the change is being internalized.
- Performance analysis phase—During this phase, the effectiveness of the OCM activities are measured.

HOW TO USE THE TOOL

Managing change has an art and science all of its own.

Ellen Florian
Author

To get started, we need to discuss some of the change management tools and approaches that are used to help ensure that the change project will be positively accepted by the people who are affected by the change. To minimize the disruptive impact that change has on the organization, a methodology called *organizational change management* (OCM) is used.

OCM should be used whenever anyone of the following conditions exists:

- When the change can have a major impact on the organization's performance
- When there is a high cost if the change is not implemented successfully
- When there is high risk that certain human factors could result in implementation failure

Change Is a Process

We can think of change truly as a process, just like any of the processes that go on within the organization. It is the process of moving from a present state or as-is state, through a transitional period that is extremely disruptive to the organization, to a future desired-state that someone believes is better than the current state (present state) (see Figure 17.1).

Change is a process

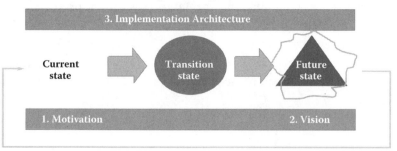

FIGURE 17.1
Change is a process.

We have to be aggressive about changing. The status quo is our enemy.

Marie Eckstein
Executive director, Dow Corning

People are very control oriented. They are the happiest and most comfortable when they know what is going to happen and their expectations are fulfilled. Keep this in mind.

Here are some important definitions to understand when discussing change management.

> **Definition:** *Change* is a condition that disrupts the current state. Change activities disrupt the current state.
>
> **Definition:** *Present state/current state/as-is state/status quo* is a state where individual expectations are being fulfilled. It is a predictable state—the normal routine.
>
> **Definition:** *Transition state* is the point in the change process where people break away from the status quo. They no longer behave as they have done in the past, yet they still have not thoroughly established the *new way* of operating. The transition state begins when the solutions disrupt individuals' expectations and they must start to change the way they work.
>
> **Definition:** *Future state* is the point where change initiatives are implemented and integrated with the behavior patterns that are required by the change. The change goals and objectives have been achieved.

It is necessary the people understand the reason for the transformation that is necessary for survival.

Dr. W. Edwards Deming
Consultant

People resist change because they are unhappy about the disruption of the current state as much or more than they are afraid of the change itself. When change occurs and expectations are not met, the four Cs (4Cs) set in:

- Competence
- Comfort
- Confidence
- Control

Change makes people feel that they are not competent enough to handle the unknown that comes along with change. Change makes people feel uncomfortable because they are entering a new world that they have not experienced before. Changes in a work environment cause people to lose their confidence. Before the change, they know their job better than anyone else, but now they have to start learning it all over again. Change causes people to feel that they have lost control over their lives and actions. From the individual's standpoint, it is the people who are making the change that are controlling their destiny. They have lost control over their lives.

When there is a disruption to the 4Cs, the emotions within the organization quickly change in a negative direction. Stress levels go up very quickly because people start to worry about what is going to happen to themselves and their friends. Productivity drops as people find more time to discuss what is going to happen to them and start to question if what they are doing is the right thing to do. The organization becomes unstable as people start to react slower to the present process. People begin to become afraid because they are uncertain about what is going to happen to them. This drives up the anxiety level of the total organization. Conflicts seem to break out everywhere. When people are worried and nervous, little things will send them into a hyper-reaction mode. Little things that would have been ignored become the most important things in life. The slightest negative comment is blown way out of proportion.

Yes, change and the accompanying disruption of expectations cause the organization to tense up, to become high strung, and explosive. The child in all of us takes over, and we become very emotional.

The focus of OCM implementation methods is on the transition from the present state to the future state. The journey from the present state to the future state can be long and perilous, and if not properly managed with appropriate strategies and tactics, it can be disastrous. Each major improvement effort will undertake this journey. This is the reason that OCM should be included in most of the project management plans.

Determining When a Change Is Major

We can apply our first *best practice* to any improvement project that is being implemented by first identifying the pitfalls. Many organizations have a tendency to assume that every change or improvement project requires the same level of implementation effort. In essence, they tend to repeat their past implementation history: they budget for cost and time requirements for both the technical and human objectives as if all change projects were the same. The best practice that should be applied here deals with accurately determining when an improvement project is going to be a major change for the people affected within the organization. If it is a major change, then it is worth some special implementation effort and some special allocation of implementation resources.

Factors to Consider

According to Daryl Conner, president and CEO of ODR Inc., there are some guidelines, in terms of when or how to determine if a major project needs special implementation effort. Essentially, there are three factors to consider:

1. Is the change a major change for the people in the organization (human impact)? A major change is any change that produces a significant disruption of an individual's normal expectation patterns. For management to determine if a change initiative is considered major, 14 specific factors should be examined. These factors that disrupt expectation patterns are
   - Amount
   - Scope
   - Transferability
   - Time
   - Predictability
   - Ability
   - Willingness
   - Values
   - Emotions
   - Knowledge
   - Behaviors
   - Logistics
   - Economics
   - Politics

   One or any combination of these factors can cause a change to be considered major in the eyes of the targets. Management must have a handle on the way their employees perceive even what seems like the most insignificant changes.

2. Is there a high cost of implementation failure? What is the price associated with failing to implement a specific improvement project (cost of failure)? It is imperative for management to understand the consequences of failing to successfully implement any change. Not only will resources be wasted on a problem that is not solved or an opportunity that is not exploited, but also there may be other implications such as morale suffering, job security being threatened, and the organization losing confidence in its leadership.

3. What are the risks that certain human factors could cause implementation failure (resistance)? Questions that need to be answered include, Is senior management truly committed to this project, how resistant will the organization be, and does this change *fit* without culture? Once again, ignoring any of these human factors can cause a project to fail. Later in this chapter, there are specific best practices that address a majority of these factors.

These three factors must be considered for every improvement project identified. Senior management must be able to recognize these business imperatives, which require a dedicated effort in managing the human and technical objectives in order to achieve success. Therefore, in order to leverage this best practice, each improvement project should be assessed with regard to these three factors. Hence, an accurate determination can be made of the level of disruption change causes the organization, and how much time, effort, money, and resources will be required to ensure successful implementation.

The flow chart in Figure 17.2 can be used to define how or if OCM should be applied to a project.

Pain Management

Change only occurs when individuals make a choice to change. We have to establish with people that there is less pain in moving forward.

William Bridges
Managing Transitions

Next, we can discuss pitfalls and best practice to building the resolve and commitment necessary, not only to initiate an improvement project but also to sustain that project all the way through to completion. One of the common pitfalls that we have seen many organizations make is strong, zealous initiation of improvement projects, only to have them flounder from

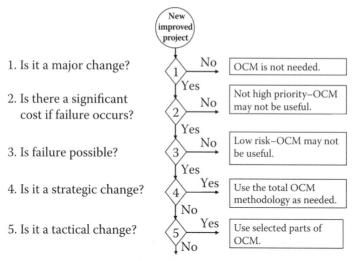

FIGURE 17.2
Applying OCM to a project.

lack of resolve to sustain the project through to completion. Obviously, then, the best practice in this case is to build the necessary commitment to sustain the change with senior and middle management, thereby enabling the organization to manage the change process over time.

Achieving *informed commitment* at the beginning of a project is one of the main issues in any change project. There is a basic formula that can be applied that addresses the perceived cost of change versus the perceived cost of maintaining the status quo. As long as people perceive the change as being more costly than maintaining the status quo, it is extremely unlikely that the resolve to sustain the change process has been built. The initiator of the change must move to increase people's perception of the high cost of maintaining the status quo and decrease their perception of the cost of the change, so that people recognize that even though the change may be expensive and frightening, maintaining the status quo is no longer viable and is, in fact, more costly. This process is referred to as *pain management*.

> When one door closes, another opens; but we often look so long and so regretfully upon the closed door that we do not see the one that has opened for us.
>
> **Alexander Graham Bell**

Pain management is the process of consciously surfacing, orchestrating, and communicating certain information in order to generate the appropriate awareness of the pain associated with maintaining the status quo compared with the pain resulting from implementing the change. The *pain* the initiator is dealing with is not actual physical pain. Rather, change-related pain refers to the level of dissatisfaction a person experiences when his or her goals are not being met or are not expected to be met because of the status quo. This pain occurs when people are paying or will pay the price for an unresolved problem or missing a key opportunity. Change-related pain can fall into one of two categories: *present-state pain* (as-is pain) and *anticipated pain* (lost-opportunity pain).

Present-state pain (as-is pain) revolves around an organization's reaction to an immediate crisis or opportunity, while anticipated pain (lost-opportunity pain), that is, the pain if the change is not made, takes a look into the future, predicting probable problems or opportunities. It is very crucial that management understands where its organization is located on this continuum of present-state pain (as-is pain) versus lost-opportunity pain (future-state pain). This understanding enables management to better time the *resolve to change*. This resolve/commitment, which must be built and sustained, can occur during either the current or anticipated time frames. If this attempt to build resolve is formed too early, it would not be sustained; if it is formed too late, it would not matter. Management has a wide variety of pain management techniques from which it can choose. Some of the techniques being used by Fortune 500 companies include cost–benefit analysis, industry benchmarking, industry trend analysis, and force-field analysis, among many others. When this process has been accepted by senior and middle management, a critical mass of pain associated with the status quo has been established, and the resolve to sustain the change process has also been established. It is only then that management can begin to manage change as a process, instead of an event.

When it comes to the employees that are affected by a change, they usually have a good understanding of the present-state pain (as-is pain) they are involved in. This may or may not be the true view of the total process' present-state pain (as-is pain). In fact, it usually does not reflect the process' real present-state pain (as-is pain). But even if the employees have a view of the process' present-state pain (as-is pain), they usually do not have any idea related to the process' anticipated pain (lost-opportunity pain). For example, if we do not make a specific change, the organization will lose 50% of its customer base and the plant may be closed, so we will all be out of work. It is for these reasons that management must communicate to all the affected employees the present state and anticipated pain related to the item being changed. It is not enough to do it one time. It must be repeated over and over again, thereby reinforcing the need for the change.

Any project that results from the performance improvement philosophy will, by necessity, cause change in an organization. The application of this best practice is critical in the beginning to mobilize support and understanding for the reasons for change, to help them let go of the status quo and move forward to a very difficult state, known as the *transition state*. Managing people through the transition state to project completion requires resolve not only to initiate change but also to sustain it over time, with management continually communicating the necessity for change and supporting the actions required to bring it about.

It is important to remember that any type of disruption stimulates resistance. The individual's existing frame of reference provides unconscious psychological security. This strong commitment with the current state can only be broken when the individual's perceived pain with the current state is greater than his or her perceived pain (fear) of the transitional and future state (see Figure 17.3).

Two factors determine the impact of the present-state pain. Each person has his or her own existing frame of reference that defines his or her level of pain related to the present state. In addition, each individual has his or her own level of pain that he or she can stand before deciding to give up on the present state and embrace the change.

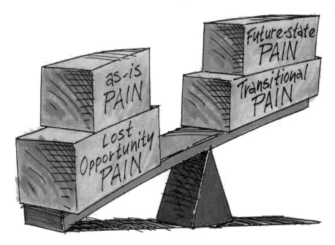

FIGURE 17.3
Pain management.

The way I see it—if you want the rainbow, you gotta put up with the rain.

Dolly Parton
Country–western singer

The first part of the equation is to generate enough pain related to the present state (as-is state) to get the individual to be open to consider making a change. The next activity is to minimize the perceived turmoil that the individual perceives he or she will go through during the transition state. This can be accomplished by providing information about the implementation plan, involving the employee in the planning, and providing education and training for the affected people early in the cycle.

The last part of the equation is to provide a realistic vision of the anticipated state (future-state) condition. We too often implement changes without taking into consideration the fabric or organizational context of the action. Contextual leadership is a key part of OCM. It is the ability to provide a specific frame of reference related to the change outputs and its impact on the affected individuals. This ability provides the people with a view of the change so that they understand what is going to happen to them and why. To provide this level of understanding, four key documents need to be prepared related to the anticipated state (future-state) solution.

- Vision statement
- Mission statement
- Operational plan
- Human attributes analysis

There's no such thing as the perfect solution. Every solution, no matter how good, creates new problems.

H. James Harrington

A vision statement should be broad enough to define the future state, yet specific enough to be personal to the affected individuals. It should be progressive enough to open everyone's eyes to what can be accomplished, but at the same time, it must be realistic. Unrealistic vision statements will get ignored as will *ho-hum* ones. A good vision statement stretches people beyond their current capabilities without setting up false hopes and dreams that cannot be obtained. The vision statement should be the stimulus that unites the team and motivates them to embrace the change.

Additionally, a good vision statement will instill enthusiasm and capture not only the individual's attention but also his or her personal desires, logic, and objectives. It is designed to bridge the gap between the individuals who are affected by the change and the organization's need for the change. The vision statement should address

- Why the organization is making the change
- How the change will affect the processes
- What is in it for the affected people and the organization
- What behavioral patterns will be affected

(In 2001) employees were ready to say, "I am going to go where you lead me because I see we need to travel in a new direction."

Marie Eckstein
Executive director, Dow Corning

The vision statement should paint a picture of the future state to the degree that the individual can judge the amount of pain that he or she will encounter when the change is implemented. All employees realize that any change will cause future-state pain, no matter how well the intentions of the change team.

The mission statement also helps the individual to better understand the future-state environment/pain. It provides an understanding of what must be accomplished if the future state is to be achieved. It will identify the human and technical objectives of the change.

The operating plan answers the question, "What is going to change?" It provides the work breakdown structure that defines how the change will be achieved.

The human attributes analysis completes the view of the future-state environment. It defines the personal attributes that will need to change in order for the change to be implemented. It will identify the need for changes in things like

- Values
- Beliefs
- Behaviors
- Attitudes
- Knowledge

With the completion and communication of these four key documents—vision statement, mission statement, operating plan, and the human attributes analysis—the individuals affected by the change will be able to make an evaluation about the degree of pain they will be subjected to when the changes are implemented. Now, based on their own experiences or management-supplied information, the affected employees should have developed in their own minds a feeling related to the pain they will or are being subjected to in the following four conditions:

- Present-state (as-is state) pain
- Lost-opportunity pain
- Transitional pain
- Future-state pain

Based on their opinion related to the pain they are experiencing and may be subjected to, they can make a decision to support the change, resist the change, or just wait and see how the change will affect them. If the combination of the present-state pain plus the lost-opportunity pain is greater than the combination of the transitional pain plus the future-state pain, the employee will support the change (see Figure 17.3). If not, they will resist it and do whatever they can to make the change project fail.

> In the case of driving for excellence in these functional areas, you have to draw attention to the crisis to really make people take notice and say, "Yes, we want to do it differently here."
>
> **Robert J. Herbold**
> *Retired chief operating officer (COO), Microsoft*

Future-State Vision Statements

> They spent their time mostly looking forward to the past.
>
> **John Osborne**
> *Look Back in Anger*

The other part of helping employees make the decision to jump is related to painting a very clear vision of the future desired state (see Figure 17.4). This vision has to address items like, What would the business processes look like? What are the technology, process, and people enablers? People need to have answers to questions like

FIGURE 17.4
Obtaining a clear vision.

- Why is this change necessary?
- What is in it for me?
- Why is it important to my organization?
- What is the downside to the change?

Once we provide them with a good understanding of the pain related to the present state and a good understanding of the future state, the employees are in the position where they can weigh in their own minds the advantages and disadvantages to them related to the change (see Figure 17.5). At

FIGURE 17.5
Balance between the as-is pain and the projected future state pain.

that point, they will make a decision to move forward or to resist your change initiative.

> Change isn't something you do by memos. You have got to involve people's bodies and souls if you want your change effort to work.
>
> **Lou Gerstner**
> *Past CEO, IBM*

Burning Platform

Daryl Conner originated the term *burning platform* to describe the activity of surfacing the pain related to the present state based on a disastrous fire that occurred on an oil-drilling platform (see Figure 17.6). Here is what Daryl Conner wrote in the book entitled *Project Change Management* (published by McGraw-Hill).

THE BURNING PLATFORM STORY

At nine-thirty on a July evening in 1988, a disastrous explosion and fire occurred on an oil-drilling platform in the North Sea off the coast of Scotland. One hundred and sixty-six crewmembers and two rescuers lost their lives in the worst catastrophe in the twenty-five year history of exporting North Sea oil. One of the sixty-three crewmembers who survived was

FIGURE 17.6
Burning platform.

a superintendent, Andy Mochan. His interview helped me find a way to describe the resolve that change winners manifest.

From the hospital bed, he told of being awakened by the explosion and alarms. He said that he ran from his quarters to the platform edge and jumped and 15 stories from the platform to the water. Because of the water's temperature, he knew that he could live a maximum of only twenty minutes if he were not rescued. Also, oil had surfaced and ignited. Yet Andy jumped 150 feet in the middle of the night into an ocean of burning oil and debris.

When asked by he took the potentially fatal leap, he did not hesitate. He said, "It was either jump or fry." He chose possible death over certain death.

Andy Mochan jumped off the oil platform into the water not because it was a good thing to do but because it was the best option he had. He jumped because the pain related to staying on the platform (present-state pain) was much greater than the pain of falling 15 stories into the water.

It is often very difficult to get people to move away from the present state that they are familiar with and take a risk that the future state will be better for them and not just better for the organization. When this is the case, management needs to define and communicate that the change is not just a good idea, but that it is a business imperative. People will accept changes if they perceive the future state will be better for them than conditions are today (see Figure 17.7).

FIGURE 17.7
People change when it is good for them.

FIGURE 17.8
Implementing change is not in line with their beliefs.

As people face change, they react in different ways. Think of boiling water as the turmoil that we face in this ever-changing world. If we have three pots of boiling water and put a carrot in one, an egg in the second, coffee grounds in the third, and let them all boil, the outcome will teach us something about how change affects different people. The carrot went in hard but came out soft and weak. The egg went in fragile but came out hard. The coffee changed the water into something better. How does this apply to you? Will you give up, become something hard, or will you transform change into triumph? As the *chef* of your own life, what do you bring to the table? (Adapted from the *Access Christian* website.)

Do not expect people to embrace change if it is not in line with their beliefs or the organization's culture (see Figure 17.8).

Change Prerequisites

It is easy to come up with new ideas, but the really hard thing to do is to give up something that worked last year.

H. James Harrington

There are three prerequisites required for change to occur and be effective:

- *Motivation*—The people who will have to change need to be motivated to accept the changes in their work environment. We will call the individuals who are the targets of the change process *change targets.*

- *Commitment*—The change targets need to understand how the change is going to affect them. This requires that the change targets be provided with a detailed understanding of how the change will affect them and their associates well before the change is implemented. This understanding of the future state is a key ingredient in creating desire to accept the change.
- *Implementation architecture*—The implementation architecture provides the bridge from the present to the future state.

Change Management Communications

You have to make it clear that change is what the goal is, what your expectation is, and it is going to be meaningful in terms of its financial and organizational impact.

Robert J. Herbold
Retired COO, Microsoft

Communications to the affected managers and employees alike must start on day one of any project and continue well after the project is complete. You need to answer the three questions in Figure 17.9 to each person over and over again.

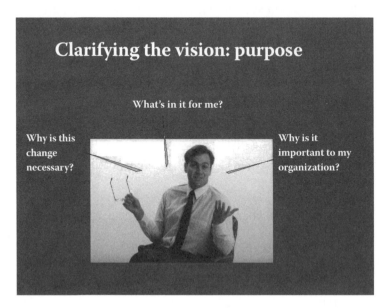

FIGURE 17.9
Why do it?

Our criminal system knows you cannot force anyone to change. All you can do is present them with the options so that they can set their own course.

Daryl R. Conner
Author

The organizational change process must create strategic communication paths to develop a rational need for change and sense of urgency related to the change. The more unpleasant the change, the more effort has to go into communicating it. If left on their own, people will blow both the bad and good news out of proportion. For example, if you are going to reengineer accounts payable with an objective of reducing cycle time, the rumor mill will soon have everyone believing that 90% of the staff will be laid off because someone heard that is what reengineering does. Change can lead very quickly to personnel problems if the communication channels are not kept open and used frequently. People seem to always discard good news when they hear bad news.

An ongoing flow of information related to the change is critical to the success of the change. Keep reminding the employees about the logic behind the change and highlight its benefits. The project team members need to be cheerleaders, focusing their discussions on how the change will benefit the organization and the staff. Management should schedule regular update meetings with all the affected staff to communicate the status of the change. Always tell the truth, even if it is unpleasant. Getting it out in the open is always better.

Bad communication builds high levels of resistance. Good communication breaks down resistance and builds resiliency. When things are changing, there always seems to be more questions than answers.

The OCM methodology needs to address two different, but similar, situations.

- How change activities are affecting the total organization
- How an individual change will influence the people it affects

Today's fast moving, rapidly changing business scene puts stress on the total organization. It seems like before a change is fully implemented, another change is affecting the first change. Today, if we really feel comfortable with the processes that we are using, they are probably obsolete.

Key Change Management Roles

> If you can put in a culture that knows change is inevitable and an opportunity, not a threat, then I think you have the potential to have a company that can grow to a very large size.
>
> **Fred Smith**
> *Founder, American Express*

In the OCM methodology, people are called upon to play five different roles. Often an individual will play different roles based on the specific situation. For example, a manager may need to change before he or she can be a sustaining sponsor. Therefore, the manager becomes a change target before he or she becomes a sustaining sponsor. The following is a list of the five different change management roles and a definition of each role.

Definition: *Initiating sponsor* is the individual or group with the power to initiate or legitimize the change for all the affected people in the organization.

Definition: *Sustaining sponsor* is the individual or group with the political, logistic, and economic proximity to the people who actually have to change. Often we talk about initiating sponsors as senior management, and sustaining sponsors as middle management, but that is not necessarily the case. Often sponsors can be someone in the organization who has no real line power but has significant influence power as a result of relationships with the people affected by the change, past successes of the individual, knowledge, or power.

Definition: *Change agent* is the individual or group with responsibility for implementing the change. They are given this responsibility by the sponsors. Agents do not have the power to legitimize change. They do not have the power to motivate the members of the organization to change, but they certainly have the responsibility for making it happen. They must depend on and leverage sponsorship when necessary.

Definition: *Change target* is the individual or group who must actually change. Many people do not like the word target. There really is nothing degrading associated with the word *target*. In fact, it is really more of an indication of where the resources, which are allocated to any specific project, must be focused to achieve successful change. If you really want to use a different name, we find *affected parties* also works well.

Definition: *Change advocate* is an individual or group who wants to achieve change but lacks sponsorship. Their role is to advise, influence, and lobby support for change.

In most cases, the project will define who will be the initiating sponsor, sustaining sponsors, and the change targets; however, you may have some choice about who will be the change agents and change advocates.

Identifying the members of an organization who must fulfill these roles, and then orchestrating them throughout the change process is a best practice that organizations can leverage to greatly increase their likelihood of success with any specific improvement project. Once these roles are identified, management should maneuver those key roles to optimize each of them throughout the change process in order to achieve successful implementation. To be effective in that task, management must understand the intricacies of each role, how they interact with each other, and how they work in an organization. The first thing that needs to be understood is that in all major change projects, key roles will overlap. When this occurs, the individual(s) should always be treated as a *target* first.

A company needs a group of top executives to be champions for this approach. If they don't develop and nurture it, they will lose the ability to manage change at scale and speed made possible by advances in technology.

Charles Kalmbach
Managing partner, Accenture

Initiating Sponsor's Role

The initiating sponsors play the role of the organization's conscience. They ensure that the proposed project will have a positive impact on the organization and that it is a priority project within the organization. They serve as checks-and-balance related to the risks that are involved in the project and its chances of success. In addition, they review the project plan, approve the project's budget, and also have the responsibility to follow the project and sit in on phase reviews. Moreover, the initiating sponsors have the power to stop a project when they believe it is not going to meet its performance objectives.

The initiating and sustaining sponsors are responsible for establishing an environment that enables the changes required by a project to be made on time and within budget. The initiating sponsor may not be the person who originates the idea; that person is the initiating advocate. Initiating sponsors do not have to ask for permission to engage in change; instead, they just keep the organization informed about what the change project is going to do, and are held accountable for its success.

Identify key people in the organization to spearhead the change. Choose them on their personal characteristics, not their place on the hierarchy. You want people who will be very strong supporters of the project and who have open minds.

Michael Tofolo
Management Review Magazine, March 1998, Buena Vista Home Video

Sustaining Sponsor's Role

Change management is a very difficult skill to master. It's easy to get up and give a speech or go to a seminar or put out a memo, but basically leading change is a day-to-day activity that takes a lot of work, a lot of energy.

Mark Huselid
Author, The HR Scorecard

The sustaining sponsor is responsible for identifying targets within their assigned span of control, understanding the target's role in the change process, and working with the targets to break down any resistance that they may have to the change. They play a major role in communicating the project's vision, mission, and objectives to the targets. They also help develop an understanding of the pain related to the present-state and anticipated future-state solutions.

Additionally, sustaining sponsors also serve by providing a continuous flow of project status to the targets and feedback to the project team about potential problems that they may encounter in dealing with the targets. They often represent the target's interest at project planning meetings.

One way to tell the difference between an initiating and sustaining sponsor is that the sustaining sponsor lives with the change after it is implemented.

Change Agent's Role

Today, change agents must be facilitators of the human aspects of change.

H. James Harrington

FIGURE 17.10
Change role.

Change agents are very special people (see Figure 17.10). They have to have excellent people skills and be able to understand and interpret their behavioral patterns. These are individuals who, until recently, were not part of most project teams. The project manager often played the role of the change agent without the required training. Basically, the change agent is responsible for making the change happen. The key attributes of an effective change agent include the following:

- Work within expectations set by the sponsor.
- Apply an in-depth understanding of how people and organizations react to the process of change.
- Value the human as well as the technical aspects of change.
- Identify, relate to, and respect the different viewpoints of sponsors, agents, and targets.
- Collect and appropriately use data regarding how and why people will resist change.
- Help build and maintain synergy among sponsors, agents, and targets.
- Communicate effectively with a broad range of people with different communication styles.
- Help build and maintain appropriate levels of commitment to the change throughout the change implementation process.
- Use his or her power and influence to achieve the goals of the change.

- Set aside personal agendas, desires, and biases that might hinder the success of the project.
- Change agents should not work harder than their sponsors.
- Change agents must be very effective and proficient at using all the change management tools.

It is very important to select excellent change agents, and they should be evaluated after each project based on the following:

- Is the individual perceived as highly credible?
- Has the individual earned the sponsor's trust and respect?
- Does the individual demonstrate a high tolerance for ambiguity?
- Does the individual thrive on challenge while avoiding stress levels associated with burnout?
- Is the individual aware of the formal and informal power structure and know how to use it?
- Does the individual have a good understanding of change management concepts and principles?
- Does the individual have a high level of political support and credibility?
- Can the individual effectively manage ambiguity and uncertainty?
- Can the individual work within the sponsor's expectations?

Individuals who have the personal traits to be change agents should be trained on the change management methodology. If the individual does not have the necessary personal traits, select another individual. It takes a long time to develop the needed personal traits (see Figure 17.11).

Preparing Change Agents in the Required Skills

We are living in a turbulent environment, where change is accelerating dramatically in three ways: volume, speed, and complexity. This means that we can no longer manage as we have in the past. A possible pitfall in this area is that sponsors, often incorrectly, assume that those identified as change agents and advocates possess the skills necessary to successfully deal with human as well as technical implementation problems.

In today's unstable environment, it is necessary to have finely honed skills for managing and implementing technical and human change. These skills are in very high demand, and special training is often necessary.

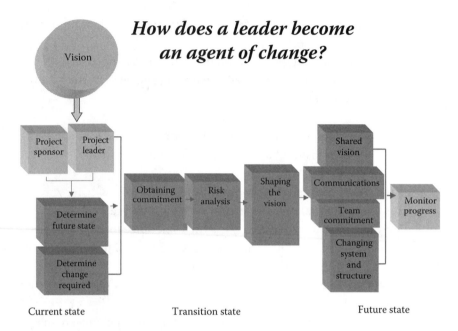

FIGURE 17.11

Transforming a leader into a change agent.

The best practice for organizations is to build the capacity to manage change. To accomplish this objective, the change agents must come into the engagement with a different perspective. Part of the development of these skills is also a bit of a shift in mind-set.

When you look at agents historically, you will note that they tended to have more of a technical expertise mentality. Their primary focus and objective was to be able to ensure that the change, whatever it might be, was technically sound and that if people were not able to use it, then it was not the agents' fault. That mind-set must shift to a mind-set that follows the idea of the facilitators of change; the mind-set that says change agents need to be responsible for not only managing the technical aspects of the change but also the human aspects of the change. This mind-set also dictates that the change agent should focus on the process as well as the content issues, with technology being designed to accommodate human interests, needs, and values.

Truly effective change agents should be skilled in a complex combination of characteristics that can be brought to bear on a given change project. Successful change agents must have the ability to work within the parameters set by the sponsor and understand the psychological dynamics regarding how individual and organizations can modify their operations.

Change agents must optimize their performance by placing emphasis on the technical aspects, and especially, the human aspects of the change. Change agents must be skilled in dealing with resistance. To understand that resistance, it is critical for change agents to be able to identify, relate to, and respect the targets' and sponsors' diverse frames of reference. Change agents must also be the *cheerleaders*. They must constantly strive to build commitment and synergy among targets and sponsors, while at the same time being aware of and utilizing power dynamics and influence techniques in a manner that reflects a capacity to achieve results in an ethical way. The bottom line is that a change agent must be professional, setting aside a personal agenda for the good of the change.

Success ultimately is judged by achievement of both the human and the technical objectives. That mind shift needs to occur if change agents are going to more effectively manage the human aspects of implementing major change projects. The change management skills within the organization are critical to the success of any change project. This change of perspective must occur if change agents are going to successfully manage the implementation of change projects. However, even the most skilled change agents in the world cannot successfully implement major change by themselves. The other roles in the change process, specifically the advocates, must have their skills finely honed and be prepared to use them.

The most important qualities in a change agent are that he or she be a visionary and a marketer—someone with good interpersonal and communication skills and someone who knows how to manage conflict. A change agent must have a spirit of inquiry. They need to be able and willing to listen to people's true concerns.

Change Agent Skills' Evaluation

The change agent skills can be evaluated using the Change Agent Skill Assessment instrument. This instrument is used to evaluate the behavior of the selected change agent. A typical evaluation leads to the following conclusions related to the project's change agents.

The analysis defines the change agents as being barriers in the following areas:

- Their understanding of the human and psychological aspects of individuals, groups, and organizational change
- Lack of ability to collect and use data to reduce resistance

- Ability to apply power and influence properly or correctly
- Ability to communicate effectively
- Lack of understanding and respect for diverse sponsor and target perspectives
- Ability to synthesize different perspectives into mutually supportive action plans

This is more typical than the usual result. Most change agents are not properly trained to do their assignment. The change agent's job is not an easy one, but it is a very important one. It is often very difficult for a technical person to perform the change agent assignment and his or her technical assignments.

Organizational Black Holes

Cascading sponsorship is an effective way to eliminate the organizational *black holes*. These are the places in the organization where change decisions enter the process, but are never heard from again. These black holes typically occur when there is a manager who does not sponsor the change, and therefore, the targets beneath him or her do not adopt the change. There is little initiating sponsors can do to maintain the change at lower levels of the organization because they do not have the logistical, economical, or political proximity to the targets. The result is that change cannot succeed if there is not a network of sustaining sponsorship that maintains the integrity of the implementation as it moves down through all levels of the organization; hence, cascading sponsorship.

The term *black hole* is referred to by astrophysicists as a location in space where gravity is so strong that all surrounding matter and energy are drown in, unable to escape. We all know people in our organization that fit that definition *to a T*. A black hole occurs whenever a sponsor stops supporting a change for any reason—be it logistic, economic, political, lack of resources, etc. It can be caused by

- Bureaucratic layers
- Cultural differences
- Geographic distance
- Personal differences
- Budgeting issues
- Lack of time

A black hole exists because the cascading sponsor did not build or sustain the needed commitment level from the individual to achieve the change objective at the local level. When a black hole exists, there are three things the relevant sponsor can do.

- Replace the uncommitted manager.
- Prepare to fail.
- Educate the managers and put them on an improvement plan. Point out to them that such behavior is not in the organization's best interest.

Building Synergy

Definition: *Synergy* is the combined action of individuals or groups working together in a manner that produces a greater total effort than the sum of their individual efforts, generates more benefits to the organization than the amount of resources consumed, promotes a higher future shock threshold, and requires less effort to change.

Definition: *Future shock* is the point at which no more change can be accommodated without the display of dysfunctional behaviors.

The next best practice is to discuss deals with the idea of synergy—building synergistic work environments and synergistic work teams. Synergy is a very important concept when implementing change projects. Synergy occurs when two or more people, working together, produce more than the sum of their individual efforts. Much has been said about empowerment, participative management, and cross-functional teams—all of which are very good ideas and necessary, but none are likely to be successful without a basic synergistic environment. A common pitfall is that management promotes the idea of synergistic output and synergistic teams, yet most fail to achieve them. The advised best practice is to enable sponsors, agents, and targets to work effectively as a synergistic team throughout the change process.

Integral to synergy is allowing people to work in a synergistic environment. Synergistic environments are open; there is no fear; there is five-way communication; and people in those environments really do feel as if they can have some influence over the outcome of any specific project or business issue. To really build this kind of environment and this kind of teamwork, it is necessary to first meet the two prerequisites of synergistic work teams:

- There needs to be a very powerful common goal shared by these sponsors, agents, targets, and advocates for the change.
- Goal achievement requires recognition of interdependency: sponsors, agents, targets, and advocates must recognize that the goal cannot be achieved without working together.

Therefore, the best practice here is primarily to focus on making sure that those prerequisites exist. Once the existing prerequisites are confirmed, a team and a process can be built so that people can really capture the potential synergy. With the prerequisites for synergy met, a group or organization can begin its journey through the four phases of the synergistic process and team development. It is important to note that all teams must go through this process—there are no shortcuts. However, the length of time a team has already been in existence can affect the duration of time it spends in each phase of the process.

These synergistic relationships are generated through a four-phase process. The four phases are as follows:

- Interacting
- Appreciative understanding
- Integrating
- Implementing

Each phase is interdependent on the others, and individuals or groups on an implementation team must demonstrate the ability and willingness to operate to the characteristics associated with each phase.

Interacting

For people to work together effectively and synergistically, they first must interact with one another. If this interaction is going to be meaningful, people must communicate effectively.

At first, this task is not easy and one usually filled with conflict. This is what is referred to as a group *storming*. What happens is that the inevitable misunderstandings, which individuals are bound to have, go unresolved. This causes anger, frustration, blaming, suspicion, alienation, hostility, and possibly withdrawal. In an attempt to stop this destructive cycle and a total breakdown of the team development process, teams must move on to group *norming*. Here, the group decides on some basic ground rules as to how it is going to operate.

Appreciative Understanding

As important as effective communication is for successful change, something more must occur. Group members in a synergistic team effort must value and utilize the diversity that exists among the members. This is a continuation of the norming process that occurs in team development. Valuing a different point of view can be difficult for individuals because of the emphasis our culture places on rational, linear, left-hemispheric thinking processes that encourage critical analyses. This thought process can produce a response and attitude similar to, "I'm right. You're wrong." However, synergy dictates that people should support each other and look for the merit in another's viewpoint.

Integrating

Even though a team has passed through the first two phases of the synergistic process, it is not yet sufficient to produce synergistic outcomes. Synergy is the result of communicating, valuing, and merging separate, diverse viewpoints. Once again, accomplishing this integration is extremely difficult because our culture does not teach and reward the skills needed. For team members to work through the norming process of team development and move on to *performing* the final phase, they need to develop the skills necessary to make integration possible. Specifically, team members need to

- Tolerate ambiguity and be persistent in the struggle for new possibilities
- Modify their own views, beliefs, and behavior to support the team
- Generate creative ways of merging diverse perspectives into new, mutually supported alternatives
- Identify issues, concepts, etc., that cannot or should not be integrated

Implementing

Even the best plans and solutions are useless unless they are fully implemented. The bottom line for synergy must be the successful implementation of change initiatives. The culmination of synergistic events should be well thought-out, change-oriented action plans. The final phase in the synergy process is designed to build on all the momentum that the previous phases have built, and to direct that energy into completing the task at hand. It is at this point in the team development process that teams actually begin to *perform*. The key to success in this phase is basic management skills. As individuals' capacity to grow beyond themselves (synergy) is increased, it must be managed as any other valuable resource would.

Most implementation problems are the result of nonsynergistic behavior. This behavior can be attributed to human nature and bad habits. Fortunately, if a team follows the guidelines developed to create synergy and effective team development, successful implementation will be the result.

Resistance to Change

> The tough part is that many times you've got to change before the real requirement to change is necessarily seen.
>
> **Art Collins**
> *CEO, Medtronics*

Resistance to change is normal and is to be expected. It is your *gut* telling you something different is taking place. In reality, it means you are on the edge of new growth. The desire for stability is part of our biological drive toward survival. In fact, if there is not some resistance to change, we would be concerned. It is a lot like antibodies attacking some organism that comes into the body. Resistance is just the natural order of things. If you are not encountering resistance, then you are not changing enough or your people just do not care. Intelligent people will always ask, "Why are we doing this?" You do not want people who follow along blindly like a flock of sheep. People's resistance to change is usually based on real and understandable concerns. On the basis of their past experiences, people believe that organizational change creates winners and losers. Some people take advantage of the change process and move ahead, while others lose power, status, and even their jobs. Psychological barriers to change are subtle. They tend to be variations of fear, like losing their familiar ways of doing things, fear of the unknown, and the fear of loss of status.

The biggest reason for a high level of resistance to change in an organization is the organization's culture, which can be described as "an invisible, impenetrable shield holding on the old behavior patterns." Organizations that are tradition bound and committed to following procedures have a hard time changing as fast as they need to. Standards like ISO 9000:2000 just add to this problem.

> The real cause of reengineering failures is not the resistance itself, but management's failure to deal with it.
>
> **Michael Hammer**
> *The Reengineering Revolution*

Definition: *Resistance to change* is defined as "any thought or action directed against a change." Some level of resistance should be expected in any major change.

Because resistance to change is inevitable, it is important that we understand why it occurs and how to minimize its impact on the change process. Too many organizations do not consider how significant the impact of resistance to change is on the organization. Too often, we believe that, if it is good for the organization, the employees will have to go along with it. The truth of the matter is that workers, back in the early part of the 20th century when we had people tied to their machines and only paid them for their physical effort, would accept change without question.

It is best for us to regard resistance as a normal necessary first step in the change process. Since resistance is inevitable, it is best that we accept it and deal with it up front, getting over this hurdle as soon as possible. This is best accomplished if the situation is addressed by a sensitive, mature, and sympathetic manager who takes the time to make each individual feel safe and illicit their true feelings and concerns. People who need to change must feel secure. Only if we have a high degree of trust in our management team will the resistance to change crumble. By involving employees in developing the individual change design and implementation plans can we often transform a person from a resistor into an advocate.

But today, we have and need people who bring more than just their physical being to work. Today's knowledge worker needs to understand why change is necessary and, when convinced that it is needed, he or she wants to actively support the change initiatives. The difference between resistance and support is very evident. With the resistant employees, their attitude is, "Let me tell you why it won't work." And with supportive employees, their attitude is, "How can I help you make it work?"

Twenty percent of the people will be against anything.

Robert F. Kennedy
U.S. attorney general

The question is—What level of resistance is normal? On the basis of our experience, the following applies:

- Thirty percent of the people will support change.
- Fifty percent of the people are undecided or neutral.
- Twenty percent of the people will resist change and often will do everything they can to make the change fail.

What it comes right down to is people have to change, or be changed. Gary Loveman, CEO of Harrah's Entertainment of Las Vegas, when discussing a change that ended the marketing fiefdoms in the different properties stated, "In the corporate marketing department I don't recall that there were more than one or two survivors." In other words, do not bother fighting change. You either fit in or get thrown out.

The Change Resistance Assessment Instrument can be used to measure

- The individual's perception of themselves as change targets
- Management's perception of their subordinates' resistance to the project
- Employees' perspective of themselves as targets

An example of one Change Resistance Assessment that was conducted on one project is as follows:

- The overall results indicated that managers perceive substantial resistance to the change project that is higher than the sponsor commitment level. This means that there is a high probability of the project not achieving its objectives on time and within budget because the target's resistance is higher than the sponsor's commitment to the project.
- It is believed that there is too little involvement of the targets in planning the change and there are inadequate rewards for accomplishing the change. There was a perception that the change was more costly than if the current state were maintained. The participants believe there are more change resistance barriers than enablers. It was determined that the change is contrary to their personal interests.

- There is a common understanding of why the change is important and necessary. There is a general trust and respect in the sponsors and agents.
- At the functional level, there is a wide degree of variation and different reasons for resisting the change.
- In one function that had the biggest differences between the sponsors and the targets, the main barriers were
  - Degree of input they had in planning the change
  - Adequacy of reward for accomplishing the change
  - Degree to which their current work patterns were considered in planning the change
  - Their needs for security
- There were no enablers noted.
- In another function that was most positive about the change, the main barriers were
  - Level of involvement in planning the change
  - Adequacy of rewards
- The enablers were
  - Compatibility with their own goals
  - Belief that their jobs will be positively affected
  - Provision of adequate resources to support the change
  - Level of respect and trust for the sponsors

It is easy to understand how having this type of information available to the project team can help them design the project in a way that greatly improves its chances of success.

In most cases, management spends far too much time reacting to the resistors because they are the squeaky wheels. They are hard to ignore. As a result, management gives them a lot of attention, most of which is wasted effort. We liken it to a small group of people outside the American Embassy protesting about something. When they get media coverage on the 6:00 PM news, they become more convinced that their cause is just.

> The conventional army loses if it does not win. The guerrilla wins if he does not lose.
>
> **Henry Kissinger**

Where management and change agents should be spending their time and effort is with the 50% of the people who are undecided. It is much easier to move them over to support the change than it is to move the resistors. Resistors use a strategy of delay. Move fast with your changes.

Resistors hate to see things move fast. You will pass by their roadblocks before they have time to set them up.

> You must be willing to let squeaky wheels squeak. Save your grease for the quieter wheels that actually are carrying the load.

Price Pritchell

Organization's Personality and Cultural Impact on Resistance to Change

The current organization's personality and culture are a huge issue that must be addressed for change projects to be implemented successfully. Because organizational personality and culture are difficult to understand, as well as hard to measure and manage, they are relatively easy to ignore. Commonly, organizations ignore them or do not treat them as key variables when implementing a major change initiative. Obviously, the best practice is just the opposite. Senior management must understand the strategic importance of the overall organizational personality and culture to the change initiative, and work hard to understand and manage the impact they have on successful implementation of improvement projects.

Corporate culture is the basic pattern of shared beliefs, behaviors, norms, values, and expectations acquired over a long period of time by members of an organization. Organizational personalities reflect the way the present management team is operating. If an improvement project or a change initiative is consistent with that set of behaviors, beliefs, norms, values, and expectations, then the organization's personality and culture are actually enablers or facilitators of that change. On the other hand, a change project may be fundamentally counter to the organization's personality and culture, making acceptance of the change much more difficult.

One thing we are very clear about is that whenever there is a discrepancy between change in culture and existing culture, the existing culture wins. So, to apply this best practice to any change initiative, we need to understand whether organizational personality and culture are enablers or barriers to the change. If they are barriers, we must identify why they are barriers, what the existing barriers are, and proactively modify the change or modify the organization's personality, or some combination of both, for change objectives to be successfully met. There are only three options available:

- Option 1: Modify the change to be more consistent with the organization's personality and culture.

- Option 2: Modify or change the organization's personality to be more consistent with the achievement of the change objectives.
- Option 3: Ignore options 1 and 2, and plan the change initiative to take significantly longer and cost significantly more than what you may have originally budgeted. (This is not really an option.)

In performing a culture assessment, 11 items need to be considered:

1. Leadership
2. Team work
3. Business
4. Structure
5. Communications
6. Knowledge flow
7. Management processes
8. Motivation
9. Decision making
10. Performance appraisal
11. Change implementation

The following are the results of a typical Culture Consistency Analysis:

- The participants classified 8 out of the 11 items as barriers or near barriers to successful implementation of the change. The most significant barriers were current structure, and incompatibility of the current status incentives and performance expectations. Only one enabler was identified: teamwork.
- There was a great deal of variation between functions. One function viewed all cultural items as barriers with the exception of teamwork that was rated at the caution level. Another function rated decision-making processes, current management process, communication, current structure, and current methods for conveying change implementations as barriers.

Building Commitment

Building commitment within all of the projects' stakeholders is essential, but few project managers seem to understand how important it is and

Phase I: Preparation
- Stage 1: Contact
- Stage 2: Awareness

Phase II: Acceptance
- Stage 3: Understanding
- Stage 4: Positive perception

Phase III: Commitment
- Stage 5: Installation
- Stage 6: Adoption
- Stage 7: Institutionalization
- Stage 8: Internalization

FIGURE 17.12
Commitment model.

know how to do it. They also do not know how easily it can be eroded. The commitment process is made up of three phases:

- Phase 1—Preparation
- Phase 2—Acceptance
- Phase 3—Commitment

Each of these three phases represents a critical juncture in the commitment process. Each phase has a number of degrees of support (stages) for the change project (see Figure 17.12).

As an individual or organization moves from one stage to the next, the commitment to the change increases. Also, the degree of effort and time required to invest in the change management process increases based on the degree of commitment required to support the change project. Figure 17.13 depicts the pluses and minuses for each stage in the commitment model.

When implementing major organizational change, there is a continuum of change management strategies that can be used depending on the change and the degree of acceptance that change must have by the targets. At one end of the commitment level is internalized commitment and at the other end is institutionalized commitment, which is forced compliance (see Figure 17.14).

Institutionalized Commitment

Not all changes require the people who are affected by the change to believe in the change. They may only be required to comply with the change. These changes that are forced upon the targets may be accepted by

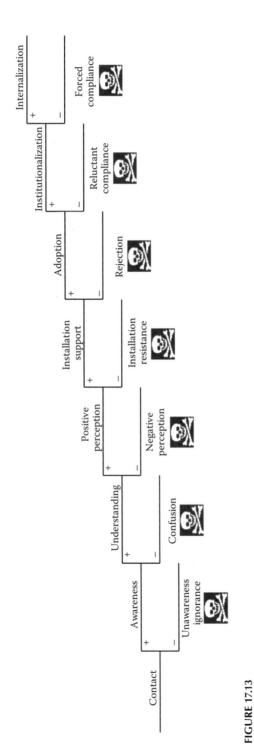

FIGURE 17.13

Stages in the commitment model: pluses and minuses.

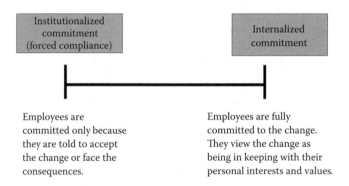

FIGURE 17.14
Range of commitment required.

the targets because they wish to comply with the organization's activities. The organization motivates the targets to comply by rewarding those who comply and punishing the individuals who do not comply. Targets often mimic acceptable behaviors and learn to do and say what they consider acceptable to the organization. Of course, this approach does not have a positive impact on the targets' attitude toward the change. In today's fast-changing environment, most organizations are realizing that its employees need to understand and support the change. With institutionalized commitment, the return on investment is often greatly reduced.

Internalized Commitment

Internalized commitment occurs when the targets believe that the change reflects their personal beliefs, needs, and wants, as well as those of the organization. This level of commitment results in the targets taking ownership for the success of the change because it satisfied their own needs and they believe it is good for the organization. At the personal level, the change is more embraced and supported than the organization could ever mandate.

Resiliency

Change is not something that just happens. The CEO and his or her key people have to make the case for change and innovation and they have to create an environment that fosters it.

Charles Kalmbach
Managing partner, Accenture

One key to survival in today's rapidly changing environment is to develop a resilient organization. Resiliency is not invented. It is liberated.

Definition: *Resiliency* is the ability to absorb high levels of disruptive change while displaying minimum dysfunctional behavior.

To increase an organization's ability to absorb change, the resiliency of the project team and those who are affected by the change (change targets) is an important factor. The more resilient the organization is, the greater its speed of change. A resilient organization has five characteristics:

- *Positive*: Resilient people display a sense of security and self-assurance that is based on their view of life as complex, but filled with opportunity. Positive individuals or groups
  - Look for the good, not the bad
  - Look forward to a better future
  - Have a high level of self-esteem
  - Feel that they can influence what is going to happen
  - Have a can-do attitude
  - Are energetic
- *Focus*: Resilient people have a clear vision of what they want to achieve. Focused individuals or groups
  - Know what they want
  - Prioritize their efforts based on impact
  - Align personal and organizational goals
- *Flexible*: Resilient people demonstrate a special ability in thinking and in working with others when responding to change. Flexible individuals or groups
  - Bend with the wind
  - Can adjust to change
  - Can see things from different perspectives
  - Are open-minded
  - Are open to other people's ideas
  - Like to be a member of a team
- *Organized*: Resilient people are able to develop and find order in ambiguity. Organized individuals or groups
  - Like structure
  - Group information effectively
  - Plan their activities
  - Are not impulsive

- *Proactive*: Resilient people encourage change, rather than defend against it. Proactive individuals or groups
  - Have lots of new ideas
  - Take risks
  - Like to see things moving along
  - Question the status quo

We liken the resilient person to a capacitor and the resistant person to a resistor. In an electronic circuit, the resistor just sets the burning-up energy; however, the capacitor stores up energy so that it can be used when needed.

The resilient person can increase his or her future shock level up to 1500 assimilation points and at the same time reduce the peak assimilation of the individual changes by as much as 50% while reducing the change impact duration by as much as 25%. This provides the organization with a very competitive advantage.

Resiliency is not a tool or a methodology; it is an attitude, a culture, the way we behave, and our beliefs. No organization can transform itself into a resilient organization overnight. It takes time to bring about the transformation. When an individual's original level of resiliency is raised through training, coaching, and rewards, it is referred to as "raising baseline resiliency to an enhanced level." As the individuals who make up the organization base resilience level move up to the enhanced level, the organization's cultures and behaviors will change to reflect this new level of resiliency.

Organizations can measure the resiliency enhancement level by monitoring the organization's changes in behavior as defined by the five resilient characteristics.

It is our experience that when a group of resilient people are affected by change, a great deal of synergy occurs. Resilient employees live with the same change challenges that everyone else had, but they usually possess the following traits:

- They are physically and emotionally healthier.
- They rebound from the change faster and with less stress.
- They achieve more of the objectives.
- They are more productive.
- They have a higher level of implementation capacity.
- They develop a resilient culture.

Winners look at change as building blocks. Those that fight change, find themselves as the foundation that someone else builds upon.

Eight Change Risk Factors

There are eight risk factors that have to be managed during any major change initiative:

1. Defining the cost of the status quo
2. Developing a clear vision
3. Obtaining sponsored commitment
4. Developing change agents and change advocacy skills
5. Understanding targeted responses
6. Aligning the change with the culture of the organization
7. Anticipating internal and external organizational events
8. Developing a sound implementation architecture

The Gartner Group estimates that inexperience, overextension, or undercommitted executive sponsorship will account for 50% of the enterprise change-initiative failures. Fewer than 35% of the change-management initiatives will include customized strategies for managing change resistors or leveraging early change adaptors unnecessarily constraining the organization's overall capacity. They also state that 75% of change leaders will employ one or more levers to help drive change without possessing even a rudimentary understanding of the implications, directly causing destructive organizational behaviors.

> Unfortunately, many organizations go for buy-in on new processes or systems after they introduced it, and the results can be catastrophic.
>
> **Robert Kritgel**
> *Consultant*

What Projects Need Change Management?

OCM is not a stand-alone project. It usually supports some other projects. It should not be applied to all projects as some progress well without the additional change efforts. Projects that should have OCM applied to them are ones with any of the following characteristics:

- All major project changes
- Projects with a high cost as a result of implementation failure
- Projects with a high risk that human factors could result in implementation failure
- Projects with an unusually short project cycle

For all the projects that meet any one of these four conditions, all the information contained in this chapter should be applied to them.

Seven Phases of the Change Management Methodology

Change management approaches may sound like common sense but, too often, common sense is not commonly practiced.

H. James Harrington

To offset the many problems that occur if the affected employees are not made part of the project before it is implemented, a seven-phase change management methodology has been developed that starts as soon as the project team is assigned (see Figure 17.15).

The following are more details related to each of the seven phases.

- Phase I—Clarify the project
  In phase I, the scope of the project and the level of commitment by management and the affected employees required for the project to succeed is defined.
- Phase II—Announce the project
  In phase II, a tailored change management plan is developed and communicated to all the affected constituents. Preplanning and sensitivity to the unique needs of various groups will minimize disruption and set the stage for acceptance of the need for the change.
- Phase III—Conduct the diagnosis
  During phase III, surveys and other types of analysis tools are used (e.g., Landscape Survey) to determine what implementation barriers exist that could jeopardize the success of the change. This diagnostic data, coupled with the rich dialogue that occurs during phase II, provides the basis for developing an effective implementation plan.
- Phase IV—Develop an implementation plan
  The implementation plan defines the activities required to successfully implement the project on time, within budget, and at an

The Seven Phases of the
Change Management Methodology

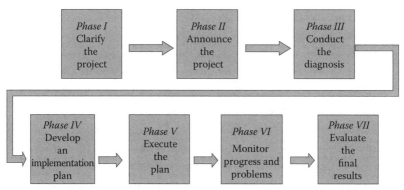

FIGURE 17.15
Seven phases of the change management methodology.

acceptable quality level. Typical things that will be addressed in this plan are

- Implications of status quo
- Implications of desired future state
- Description of the change
- Outcome measures
- Burning platform criteria
- Comprehensive or select application of implementation architecture
- Disruption to the organization
- Barriers to implementation
- Primary sponsors, change agents, change targets, and advocates
- Tailoring of announcement for each constituency
- Approach to pain management strategies
- Actions to disconfirm status quo
- Alignment of rhetoric and consequence management structure
- Management of transition state
- Level of commitment needed from which people
- Alignment of project and culture
- Strategies to improve synergy
- Training for key people
- Tactical action steps
- Major activities
- Sequence of events

- Phase V—Execute the plan

  The goal of phase V is to fully achieve the human and technical objectives of the change project on time and within schedule. It is designed to achieve these objectives by reducing resistance and increasing commitment to the project.

- Phase VI—Monitor progress and problems

  The goal of phase VI is to keep project implementation on track by consistently monitoring results against plan.

- Phase VII—Evaluate the final results

  The intent of phase VII is to provide a systematic and objective collection of data to determine if the tangible and intangible objectives of the project have been achieved and to provide insight into lessons learned and potential problem areas that may arise in future change projects.

Change Management Tools

Change management includes almost 50 unique change management tools. Some of them are

- Cultural assessment
- Landscape surveys
- Change agent evaluation
- Change history survey
- Change resistance scale
- Overload index
- Predicting the impact of change
- Role map application tool
- When to apply implementation architecture

Table 17.1 indicates where these tools can be used in each of the seven phases of the change management methodology. The book *Project Change Management*, written by H. James Harrington, Daryl R. Conner, and Nicholas L. Horney (2000), provides detailed information on each of these 50 unique change management tools. The surveys and evaluations that make up the 50 unique change management tools were developed by Daryl R. Conner and his team at O.D.R. (O.D.R. is now called Conner Partners.)

TABLE 17.1

Change Management Tool versus the Project That They Are Used In

OCM Assessments, Planning, Tools, and Training	Pre-Work[t]	Phases						
		I	II	III	IV	V	VI	VII
Change agent evaluation (A)	×			×				
Change agent selection form (A)	×			×				
Change history survey (A)[a]		×						
Change project description form (P)	×	×	×	×	×	×	×	×
Change resistance scale (A)				×				
Communicating change: project analysis (P)			×					
Communicating change: constituency analysis (P)			×					
Communicating change: statement development (P)			×					
Communicating change: announcement plan (P)			×					
Culture assessment (A)				×				
Culture audit (A)				×				
Expectations for a successful change project (A)		×						
Implementation plan advocacy kit (P)					×			
Implementation plan evaluation (A)					×			
Implementation problems assessment (A)				×				
Landscape survey (A)[a]		×		×		×	×	×
OCM training for sponsors, agents, targets, and advocates (T)	×	×			×			
Organizational change implementation plan (P)					×	×	×	×
Overload index (A)[a]		×				×		
Pain management strategies: sponsor (P)	×							
Post mortem process[b]								×
Predicting the impact of change (A)		×		×				
Preliminary implementation plan (P)					×			
Role map application tool (P)	×	×	×	×	×	×	×	×
Senior team value for discipline (A)		×						
Sponsor checklist (A)		×		×				
Sponsor evaluation (A)		×		×				
Synergy survey (A)		×		×				
When to apply implementation architecture (A)		×						

Note: Pre-Work[t]: used preliminarily to starting phase I.

[a] This assessment tool is scored by ODR's Diagnostic Services.

[b] This project-effectiveness evaluation tool is not OCM specific.

These tools can be purchased by contacting Conner Partners (www.conner partners.com), or by mail to 1230 Peachtree Street, Suite 1000, Atlanta, Georgia, 30309 (phone: 404 564-4800).

EXAMPLES

See factors to be considered in the "How to Use the Tool" section of this chapter.

SOFTWARE

Some commercial software available includes but is not limited to

- Intelex Management of Change: www.intelex.com
- Change Management Enterprise Software: www.changemanagement .net
- Techwatch: www.jazdtech.com

REFERENCE

Harrington, H.J., Conner, D.R., and Horney, N.L. *Project Change Management.* New York: McGraw-Hill, 2000.

SUGGESTED ADDITIONAL READING

Conner, D.R. *Managing at the Speed of Change.* New York: Random House, 1993.
Harrington, H.J. *Change Management Excellence.* Chico, CA: Paton Press LLC, 2006.

18

Potential Investor Presentations

H. James Harrington

CONTENTS

Selling requires both parties participating.

H. James Harrington

DEFINITION

Potential investor presentation is a short PowerPoint presentation designed to convince an individual or group to invest their money in an organization or a potential project. It can be a presentation to an individual or group not part of the organization or the management of the organization that the presenter is presently employed by. It is usually part of a short meeting that usually lasts no more than 1 hour.

USER

This tool can be used by individuals or small groups of people from the same organization.

OFTEN USED IN THE FOLLOWING PHASES OF THE INNOVATIVE PROCESS

The following are the seven phases of the innovative cycle. An X after the phase name indicates that the tool/methodology is used during that specific phase.

- Creation phase
- Value proposition phase
- Resourcing phase X
- Documentation phase
- Production phase
- Sales/delivery phase
- Performance analysis phase

TOOL ACTIVITY BY PHASE

- Resourcing phase—During this phase, it is used to convince an individual, group, or organization to invest resources in an organization or a project.

HOW TO USE THE TOOL

Acquiring Investment from Outside the Organization

When making a presentation to an outside investor, the presentation needs to sell not only the product or output from the organization but also the caliber of the people that make up the organization, including its board of directors. Typically, these meetings with potential investors are scheduled for 1 hour. Be careful not to use up the hour making the presentation so that the potential investor does not have time to say yes. All too often, the presenters are so anxious to sell the product and services that they do not leave time for discussion. A good rule of thumb is to limit the presentation to no more than 75% of the meeting scheduled time. (And 50% is even better.) All too often, a presenter will try to answer all of the potential questions the investor will have as part of the presentation. On occasion, I have purposely left some key points out of the presentation that the investor is almost sure to ask since that stimulates the conversation and discussion part of the meeting. This does not mean that you should not be prepared to answer any potential investor's question. In fact, you should make the presentation a number of times to different friends, asking them to cross-examine you with as many questions as they can think of. Nothing leaves a better impression than being able to answer the potential investors' questions without hesitation and with all the facts and figures.

Often we think of investors only in terms of investing money, but often we are approaching them for a great deal more. For example, we may want their personal time, connections, equipment, office space, distribution system, etc. Remember, you are approaching potential investors asking for the use of their resources, not just their money. Keep this in mind and continuously focus on the benefits. Sell how they will benefit from allowing you to use the resources. I particularly have selected the word *use* because even if it is money you are asking for, the investor is expecting to get it back with a significant return on their investment.

The following are some helpful hints related to making a compelling pitch to help hook the investor.

- Point out what advantages the opportunity or product has as frequently as possible.
- Point out frequently what the investor will get as a result of their investment. They are looking for terms like return on investment, increased stock value, acquisition by larger organization, percent growth, profitability, growing market, etc.
- Avoid font size less than 20.
- Do not read the bullet points—have a conversation.
- Never used dark colors on dark backgrounds or light colors on light backgrounds. Try not to use combinations like red/green, brown/

gray, and blue/purple. Remember that some of your investors may be color-blind.

- Design a standard PowerPoint so that it reflects the organization's culture; that is, use the company's colors and logo as part of the basic presentation layout.
- Use graphs, charts, and drawings whenever you can to replace text when possible. Remember that a picture is worth a thousand words.
- Start with a strong catchy introduction, and end with even a stronger one to close the presentation.
- Do your homework and be sure you know the audience—their interests and where they have invested in the past.
- No more than two people should make the presentation.
- Be enthusiastic and passionate about the organization.
- You cannot practice too much prior to making the presentation.

Think about it for a minute. What does the potential investor want to know? An investor first wants an overview that tells him why he should spend his time with you. This means answering questions like

- Who are you and what role do you play in your organization?
- What is the product or service the organization provides?
- How long has the organization been in existence?
- How much money are you trying to raise? And how much money have you raised so far?
- What are you going to do with the money raised?
- How much money have the owners of the organization invested?

An investor wants to know why the output for the organization is needed. What is the problem you are trying to solve? Is your organization qualified to solve that problem? This means being able to answer questions like

- What is the output from the organization?
- What need is it fulfilling?
- How much is the output needed?
- What other organizations are creating similar outputs?
- What will be unique about your output?
- How big is the market you are trying to service, and what percentage of that market will your organization acquire?

The potential investor will want to gain an understanding of the key individuals within the organization skills/experience/competencies related to organizing and running an organization. Board members and existing investors add credibility and references, so be sure to discuss them. The investor will be looking for the following types of information.

- Who are the key officers and technical staff and what is their previous experience? Be sure to highlight any venture-backed experiences with successful sale/initial public offering. Do not be afraid to talk about your failures. In Silicon Valley, you are not considered an entrepreneur until you have had two failures. Be sure to highlight what you have learned from your failures. Use words like *I should have, I shouldn't have, I didn't realize, I learned to,* etc.
- How many full-time employees are on board and what is your estimated growth needs?
- How successful have the individuals who are presently invested in the organization been on investing in successful organizations?
- Who are on the board of directors? How successful have the organizations they have managed been? How involved are they in your organization's operations?

The investor will want to have information related to what the organization is going to provide to the external customers. They want to know about the technology and how the output will be created. It is wise to focus more on what the output does and why the customer will procure it rather than on how it is created. It is best to focus on the benefits that will be derived rather than the technology that produces the output. In some cases, the customer will want to know more about the technology, but let them show a real interest by asking questions. The investor will want to know what stage the process is at that is creating the output. Be sure they understand what the primary output does and how the customer will benefit from it. For advanced technology, be sure you explain why it will perform as projected. Using pictures of the output in use is always a good approach.

Investors also want to have a good understanding of the market the organization will be serving.

- Who are the customers? Are you servicing any of them today?
- What is your pricing model?

- How you market your output to your customers?
- What is your distribution strategy?
- How do you plan to get to the market faster and cheaper than the competition?
- Who will you be competing against?
- How does their product differ from yours?
- At what stage of development is the competitor's product?

Of course, potential investors will always want to have a good understanding of the financial status of the organization and its projected financial future. Investors want to know when the breakeven point will be reached and when they will start making money through their investment. You need to answer typical questions like the following:

- How much money has been invested so far?
- How much is the organization in debt?
- What revenue has been generated to date?
- What do the year-end projections look like? What are the projections for next year and for a 5-year period?
- What are the top-line revenue, cost, and margin figures?
- When is the breakeven point projected to be reached?
- How long will the funds last?
- Will you be looking for additional rounds of funding?

Potential investors will want to see the output development schedule and a financial expenditure and income projections. Typically, a Gantt chart is an excellent way of showing this.

Make sure you have a very strong summary that highlights the key points that you have made during the presentation. Assume they can make a decision in a very short time. Encourage discussion throughout the meeting. Before the end of the meeting, ask them for their investment. If they are not ready to make that decision, ask them when the decision will be made. If they cannot make a decision now, try to establish a date for second meeting. Do not give up because they do not decide to invest in your organization/products. Both IBM and HP turned down the photocopier forcing the entrepreneur to go up and start what today is known as Xerox.

Potential Internal Investors

Every time you have a good idea and you want to have it implemented, you will need to have the organization invest its limited resources to make it happen. When you ask the organization to take on a new project or even to make a simple change to the way a process is being conducted, you are asking the organization to invest time, people, space, and money that could have been directed toward other endeavors. Keep this in mind as you prepare to make a presentation to your management team asking them to support a new product or a change to a product, process, or the organizational structure.

There are a lot of things in common between a presentation to an external investor and the presentation to your own organization's management team where you are asking them to apply resources to take advantage of an opportunity. In both cases, the individuals are managing a limited set of resources that they have to maximize how effectively they are utilized. In both cases, the presenter is faced with a limited amount of time to convince these key individuals that investing in your particular opportunity is the best way for their organization to utilize its resources. In both cases, you need to convince them that you and your team have the capabilities of accomplishing its objectives at an acceptable risk level. They also need to understand the proposed activity and a supporting technology well enough to give them assurance that the endeavor has a reasonable chance of being completed successfully.

Think about it for a minute. What does the potential internal investor want to know? An internal investor first wants an overview that tells him why he should spend his time with you discussing the subject. This means answering questions like

- Who are you and what role do you play in your organization?
- Why are you there?
- What is the value-added advantage of this opportunity for the organization?
- How does this impact present operations?
- What is the output from the organization?
- What need is it fulfilling?
- How much is the output needed?
- What other organizations are creating similar outputs?
- What will be unique about your output?

- How big is the market you are trying to service, and what percentage of that market will your organization acquire?
- How will it fit into the organization's culture and strategic plan?

The internal investor will want to have information related to what the organization is going to provide to the external customers. They want to know about the technology and how the output will be created. The internal investor is typically more interested in the technology and how it will be applied than the external investor. The internal investor will be more interested in how the present equipment and technology can be reapplied to minimize the impact it will have on the organization. They will want to know what stage the process is in that creates the output. Be sure they understand what the primary output does and how the customer will benefit from it. For advanced technology, be sure you explain why it will perform as projected. Pictures of the output in use are always a good approach.

Internal investors also want to have a good understanding of the market the organization will be serving.

- Who are the customers? Are you servicing any of them today?
- What is your pricing model?
- How will the marketing and distribution strategy make use of today's processes?
- How do you plan to get to the market faster and cheaper than the competition?
- Who will you be competing against?
- How does their product differ from yours?
- At what stage of development is the competitor's product?

Of course, potential internal investors will always want to have a good understanding of how taking advantage of this opportunity will influence the financial future of the organization. Internal investors want to know when the breakeven point will be reached and when they will start making money on their investment. You need to answer typical questions like the following:

- How much money has been invested so far?
- What revenue has been generated to date?

- What does the year-end projection look like? What are the projections for next year and for a 5-year period?
- What are the top-line revenue, cost, and margin figures?
- When is the breakeven point projected to be reached?
- What other resources are required and how much of them as needed?

Internal investors will want to see the output development schedule and a financial expenditure and income projections. Typically, a Gantt chart is an excellent way of showing this.

It does not make any difference if you are presenting to an internal or an external investor. You need to have a very strong summary that highlights the key points that you have been making during the presentation. Encourage discussion as it often helps the investor to understand and buy into the project. In utilizing the time you have with the investor, be sure that you give them time to make a decision and to agree to provide the resources. One big mistake a presenter often makes is not to ask the investor if they are going to provide the resources. If they cannot make a decision during the meeting, try to establish a date when the decision will be made. If they cannot make a decision, offer to provide them with any additional information they would need to make the decision, or volunteer to come back at a later date to answer any questions they may have.

EXAMPLES

To keep the example pages to a minimum, we are just going to provide a typical narrative that would be included on a typical presentation from a company called 2 XL.

PowerPoint number one:
2 XL Inc.
Makers of Advanced Project Management Software
Presenter: H. James Harrington, chief executive officer (CEO)
PowerPoint number two:
Information Related to 2 XL Inc.
- 2 XL is a software development and distribution company that produces software that combines project management,

knowledge management, and resource management activities together.
- We have been in existence since 2010.
- Presently, we have a staff of 45 programmers, 10 salesmen, and 20 supporting staff.

PowerPoint number three:

Reason for Additional Funding
- Present product is designed for small to midsize companies with a total staff of 800 or fewer.
- We require $6 million to rewrite the software programs to expand capabilities to cover large companies with a total staff of under 200,000.
- Our present software capability covers only 800 employees, and it has resulted in the loss of potential sales of $4 million during the last 12 months.

PowerPoint number four:

Market Analysis
- With increased emphasis on innovation, the present market value of $1.8 billion is projected to grow to $3.2 billion by 2018.
- Gartner has evaluated our product in the lower right-hand quadrant. We were rated the highest in capabilities but very low in number of installations.
- Our primary competitor has been IBM. Gartners' evaluation of IBM's capabilities is being 30% lower than ours. But IBM's number of installations is 90% higher than ours. This puts IBM in the very positive upper right-hand quadrant of the analysis, while we were in the putting them in the lower right-hand quadrant of their analysis.
- Expanding our service capabilities will allow us to compete favorably in the high-end market where the maximum profit per sale is realized.

PowerPoint number five:

Product Information
- The product combines the databases for project management and resource management.
- This allows for accurate tracking of resource utilization and identification of available resources.
- It quickly defines a match between needed skills and present skills available.

- It automatically adjusts work assignments on the basis of changes in schedules and priorities.
- It captures projected and actual resource projections and utilization, allowing it to serve as a base for the program plan of future similar-type initiatives.
- It captures best practices to allow them to be applied to future initiatives.

PowerPoint number six:

Customer Application

- Inputs to the software package (presenter describe various inputs required—should be part of a demonstration)
- Outputs from the software package (presenter describes typical outputs—should be part of a demonstration)
- How customer uses the outputs
- The value added to the customer of the major outputs

PowerPoint number seven:

Goals and Objectives

- (A Gantt chart is best used here with key milestones highlighted and accomplishments at each milestone discussed.)

PowerPoint number eight:

Financial Plan

- Investment in the present software package $8 million.
- Product cost per user to be reduced by 10% for applications that exceed 1000 employees.
- Presently, we have $2 million committed pending raising the additional $4 million.
- The net profit last year was $250,000. This year is estimated to exceed $650,000.
- The new investment funds are projected to be approximately used for $5 million in programming expenses and $1 million in marketing expenses.

2 XL Team

2 XL Officer

- CEO Dr. H. James Harrington—10 years principal with Ernst & Young
- COO Dr. Frank Voehl—previously executive with Florida Power and Light
- CFO Candy Rogers—previously chief financial officer for Harrington Institute

- VP R&D—Charles Mignosa—past director of research at IBM
- CIO—Jim Sylvester—previously with Ernst & Young

2 XL Board of Directors

- Chairman: Dr. John White—CEO Stratford International
- Kenneth Lomax—vice president, International Marketing
- Dr. Nita Tripp—software development manager, Pearson Institute
- Dr. M. Heidi—technical advisor to Chamber of Commerce
- Richard Jackson—vice president of Inwell Bank

Summary of Opportunity

- The technical level of the present and proposed products is superior to that of the competition.
- The market for the product is expanding rapidly.
- The basic technology has been defined and piloted.
- The breakeven point is 18 months.
- We have a very strong and experienced management team and board of directors.
- Present users' satisfaction level is more than 90%.
- There are no known barriers to entering the market.
- Risks related to meeting objectives are minimal with no known significant obstacles to overcome.

SOFTWARE

Some commercial software available includes but is not limited to

- Investment Mgmt Software: www.Capterra.com
- MediaShout 5: MediaShout.com
- SideGenius: sidegenius.com

SUGGESTED ADDITIONAL READING

Gallo, C. *The Presentation Secrets of Steve Jobs: How to Be Insanely Great in Front of Any Audience*. New York: McGraw-Hill, 2010.

Harrington, H.J. *Resource Management Excellence.* Chico, CA: Paton Press, 2007.

Harrington, H.J. and Harrington, J.S. *Total Improvement Management.* New York: McGraw-Hill, 1995.

Reynolds, G. *Presentation Zen: Simple Ideas on Presentation Design and Delivery.* Berkeley, CA: New Riders, 2012.

19

Project Management (PM)

H. James Harrington

CONTENTS

DEFINITION

Project management is the application of knowledge, skills, tools, and techniques to project activities in order to meet or exceed stakeholders' needs and expectations from a project. (*Source:* Project Management Body of Knowledge [PMBOK] Guide.)

USERS

Project management is most effectively used with groups and teams across different functions. Key members from each function involved in a project should be part of the project management team.

OFTEN USED IN THE FOLLOWING PHASES OF THE INNOVATIVE PROCESS

The following are the seven phases of the innovative cycle. An X after the phase name indicates that the tool/methodology is used during that specific phase.

- Creation phase X
- Value proposition phase X
- Resourcing phase X
- Documentation phase X
- Production phase X
- Sales/delivery phase X
- Performance analysis phase X

TOOL ACTIVITY BY PHASE

- All phases—Project management is used in all phases of the innovation process. In some initiatives, a project is formed as soon as a potential unfulfilled need is identified. By the time the value proposition phase is completed, a formal project plan with all of its elements is often documented and approved. The remaining five phases use the project plan as an outline that defines what, when, and how activity should be completed.

HOW THE TOOL IS USED

Introduction

A project plan should be based on and be supported by two other major plans:

Business plan (BP): A BP is a formal statement of a set of business goals, the reason why they are believed to be attainable, the resources required, and the plan for reaching these goals. It also contains background information about the organization or team attempting to reach these goals.

Annual operating plan (AOP): The AOP is a formal statement of business short-range goals, and the reason they are believed to be attainable, the plan for reaching these goals, and the funding approved for each part of the organization (budget). It is included in the implementation plan for the coming (years one through three) of the strategic plan. It may also contain background information about the organization or team attempting to reach these goals. One of the end results is performance planning for each manager and employee who will be implementing the plan over the coming year. The AOP is often just referred to as the operating plan (OP).

Performance improvement occurs mainly from a number of large and small projects that are undertaken by the organization. These projects involve all levels within the organization and can take less than a hundred hours or millions of hours to complete. They are a critical part of the way an organization's business strategies are implemented. It is extremely important that these multitudes of projects are managed effectively if the stakeholder's needs and expectations are to be met. This is made more complex because conflicting demands are often placed on the project. For example:

- Scope
- Time
- Cost
- Quality
- Stakeholders with different identified (needs) and unidentified (expectations) requirements

The different stakeholders often place conflicting requirements on a single project. For example, management wants the project to reduce labor cost by 80% and organized labor wants it to create more jobs.

The Project Management Institute in Upper Danbury, Pennsylvania, is the leader in defining the body of knowledge for project management. Their PMBOK approach to project management has been widely accepted throughout the world. In addition, the International Organization for Standardization's Technical Committee 176 has released an international standard ISO/DIS 10006: Guidelines to Quality in Project Management. These two methodologies complement each other and march hand in hand with each other.

- Project: A temporary endeavor undertaken to create a unique product or service.
- Program: A group of related projects managed in a coordinated way. Programs usually include an element of ongoing activities.

Large projects are often managed by professional project managers who have no other assignments. However, in most organizations, individuals who serve as project manager only have a small percentage of their time allocated to manage many projects. In either case, the individual project manager is responsible for defining a process by which a project is initiated, controlled, and brought to a successful conclusion. This requires the following:

- Project completed on time
- Project completed in budget
- Outputs met specification

- Customers are satisfied
- Team members gain satisfaction as a result of the project

A good project manager follows General George S. Patton's advice when he said, "Don't tell soldiers how to do something. Tell them what to do and you will be amazed at their ingenuity." Although a single project life cycle is very difficult to get everyone to agree to, the project life cycle defined in U.S. Department of Defense (DOD) document 5000.2 (revision 2-26-92 entitled "Representative Life Cycle for Defense Acquisition") provides a reasonably good starting point. It is divided into five phases (see below).

- Phase 0—Concept exploration and definition
- Phase I—Demonstration and validation
- Phase II—Engineering and manufacturing development
- Phase III—Production and deployment
- Phase IV—Operations and support

A life cycle that we like better for an organization that is providing a product and service is shown below and in Figure 19.1.

- Phase I—Concept and definition
- Phase II—Design and development
- Phase III—Creating the product or service
- Phase IV—Installation
- Phase V—Operating and maintenance
- Phase VI—Disposal

Traditionally, projects have followed a pattern of phases from concept to termination. Each phase has particular characteristics that distinguish it from the other phases. Each phase forms part of a logical sequence in which the fundamental and technical specification of the end product or service is progressively defined.

The successful project manager understands that there are four key factors that have to be considered when the project plan is developed. All four factors overlap to a degree, but should first be considered independently and then altogether (see Figure 19.2).

To effectively manage a project, the individual assigned will be required to address the 11 elements defined.

PHASE					
I	II	III	IV	V	VI
Concept and design	Design and development	Manufacturing	Installation	Operation and maintenance	Disposal

• New product opportunities • Analysis of system concept and options • Product selection • Technology selection • Make/buy decisions • Identify cost drivers • Construction assessment • Manufacturability assessments • Warranty incentives	• Design trade-offs • Source selection • Configurations and change controls • Test strategies • Repair/throwaway decisions • Performance tailoring • Support strategies • New product introduction	• System integration and verification • Cost avoidance/cost reduction benefits • Operating and maintenance cost monitoring • Product modifications and service enhancements • Maintenance support resource allocation and optimization	• Retirement cost impact • Replacement/renewal schemes • Disposal and salvage value

FIGURE 19.1
Sample applications of project life cycle (PLC).

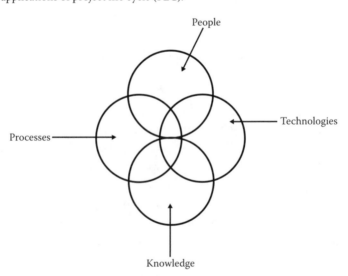

FIGURE 19.2
Key program management factors.

Knowledge Areas

The 11 knowledge areas are as follows:

1. *Project Integration Management*: Project Integration Management includes the processes and activities needed to identify, define, combine, unify, and coordinate the various processes and project management activities within the project management process groups.
2. *Project Scope Management*: Project Scope Management includes the processes required to ensure that the project includes all the work required, and only the work required, to complete the project successfully.
3. *Project Time Management*: Project Time Management includes the processes required to manage the timely completion of the project.
4. *Project Cost Management*: Project Cost Management includes the processes involved in planning, estimating, budgeting, financing, funding, managing, and controlling costs so that the project can be completed within the approved budget.
5. *Project Quality Management*: Project Quality Management includes the processes and activities of the performing organization that determine quality policies, objectives, and responsibilities so that the project will satisfy the needs for which it was undertaken.
6. *Project Human Resource Management*: Project Human Resource Management includes the processes that organize, manage, and lead the project team.
7. *Project Communications Management*: Project Communications Management includes the processes that are required to ensure timely and appropriate planning, collection, creation, distribution, storage, retrieval, management, control, monitoring, and the ultimate disposition of project information.
8. *Project Risk Management*: Project Risk Management includes the processes of conducting risk management planning, identification, analysis, response planning, and controlling risk on a project.
9. *Project Procurement Management*: Project Procurement Management includes the processes necessary to purchase or acquire products, services, or results needed from outside the project team.
10. *Project Stakeholders Management*: Project Stakeholder Management includes the processes required to identify all people or organizations affected by the project, analyzing stakeholder expectations

and impact on the project, and developing appropriate management strategies for effectively engaging stakeholders in project decisions and execution.

11. *Project Information Management*: Project Information Management includes the activities required to understand a project's requirements related to the handling, analysis, and reporting of related data and the capturing of knowledge related to the project performance.

They do not include organizational change management (OCM) as a separate element but it is placed under the element entitled *project risk management*. Of course, the depth and detail that each element needs in order to be evaluated and managed will vary greatly depending on the scope and complexity of the project. I personally feel that OCM plays such a critical role in successfully implementing a high percentage of the projects that it should be included as the 12th element.

The book *A Guide to the Project Management Body of Knowledge* published by the Project Management Institute summarizes the project management knowledge areas as follows: (Note: We have added a 12th knowledge area entitled Project Change Management that is not included in the Project Management Institute's body of knowledge.)

Project Integration Management

A subset of project management that includes the processes required to ensure that the various elements of the project are properly coordinated. It consists of

- Project plan development—taking the results of other planning processes and putting them into a consistent, coherent document.
- Project plan execution—carrying out the project plan by performing the activities included therein.
- Overall change control—coordinating changes across the entire project.

Project Scope Management

A subset of project management that includes the processes required to ensure that the project includes all the work required, and only the work required, to complete the project successfully. It consists of

- Initiation—committing the organization to begin the next phase of the project.
- Scope planning—developing a written scope statement as the basis for future project decisions.
- Scope definition—subdividing the major project deliverables into smaller, more manageable components.
- Scope verification—formalizing acceptance of the project scope.
- Scope change control—controlling changes to project scope.

Project Time Management

A subset of project management that includes the processes required to ensure timely completion of the project. It consists of

- Activity definition—identifying the specific activities that must be performed to produce the various project deliverables.
- Activity sequencing—identifying and documenting interactivity dependencies.
- Activity duration estimating—estimating the number of work periods that will be needed to complete individual activities.
- Schedule development—analyzing activity sequences, activity durations, and resource requirements to create the project schedule.
- Schedule control—controlling changes to the project schedule.

Project Cost Management

A subset of project management that includes the processes required to ensure that the project is completed within the approved budget. It consists of

- Resource planning—determining what resources (people, equipment, materials) and what quantities of each should be used to perform project activities.
- Cost estimating—developing an approximation (estimate) of the costs of the resources needed to complete project activities.
- Cost budgeting—allocating the overall cost estimate to individual work items.
- Cost control—controlling changes to the project budget.

Project Quality Management

A subset of project management that includes the processes required to ensure that the project will satisfy the needs for which it was undertaken. It consists of

- Quality planning—identifying which quality standards are relevant to the project and determining how to satisfy them.
- Quality assurance—evaluating overall project performance on a regular basis to provide confidence that the project will satisfy the relevant quality standards.
- Quality control—monitoring specific project results to determine if they comply with relevant quality standards and identifying ways to eliminate causes of unsatisfactory performance.

Project Human Resource Management

A subset of project management that includes the processes required to make the most effective use of the people involved with the project. It consists of

- Organizational planning—identifying, documenting, and assigning project roles, responsibilities, and reporting relationships.
- Staff acquisition—getting the human resources needed assigned to and working on the project.
- Team development—developing individual and group skills to enhance project performance.

Project Communications Management

A subset of project management that includes the processes required to ensure timely and appropriate generation, collection, dissemination, storage, and ultimate disposition of project information. It consists of

- Communications planning—determining the information and communications needs of the stakeholders: who needs what information, when will they need it, and how will it be given to them.

- Information distribution—making needed information available to project stakeholders in a timely manner.
- Performance reporting—collecting and disseminating performance information. This includes status reporting, progress measurement, and forecasting.
- Administrative closure—generating, gathering, and disseminating information to formalize phase or project completion.

Project Risk Management

A subset of project management that includes the processes concerned with identifying, analyzing, and responding to project risk. It consists of

- Risk identification—determining which risks are likely to affect the project and documenting the characteristics of each.
- Risk quantification—evaluating risks and risk interactions to assess the range of possible project outcomes.
- Risk response development—defining enhancement steps for opportunities and responses to threats.
- Risk response control—responding to changes in risk over the course of the project.

Project Procurement Management

A subset of project management that includes the processes required to acquire goods and services from outside the performing organization. It consists of

- Procurement planning—determining what to procure and when.
- Solicitation planning—documenting product requirements and identifying potential sources.
- Solicitation—obtaining quotations, bids, offers, or proposals as appropriate.
- Source selection—choosing from among potential sellers.
- Contract administration—managing the relationship with the seller.
- Contract closeout—completion and settlement of the contract, including resolution of any open items.

Project Information Management

Effective management of data, information, and knowledge is becoming increasingly important in order to support today's advanced projects and systems. Data collection systems are now available that easily and automatically collect relevant data and instantaneously report it back to individuals that can react to any negative changes and evaluate the effectiveness of changes as they are implemented. A subset of project information management consists of

- Management information systems planning—identifying the project requirements for the use of data, information, and the capturing of knowledge, then defining how these needs can be accommodated.
- Management information systems implementation—the acquiring or developing of the required software packages, and obtaining the required equipment.
- Management information systems training—training the information systems and using personnel on the specific information technology that will be used in support of the project.
- Documentation—documenting the management information system so that it can be effectively maintained.
- Reporting—providing analyzed information back to the users of the management information system output so they can manage, control, and improve the project and its outputs.
- Integration—integrating the individual project's management information system into the total quality information system used throughout the organization.

Organizational Change Management (OCM)

This is a part of project management that is directed at the people side of the project. OCM helps prepare the people who either live in the process that is being changed or have their work lives changed as a result of the project not to resist the change.

OCM often prepares the employees so well that they look forward to the change. (Note: This is not part of the PMBK project management concept).

- OCM planning—Define the level of resistance to change and prepare a plan to offset the resistance.

- Define roles and develop competencies—Identify who will serve as sponsors, change agents, change targets, and change advocates. Then, train each individual on how to perform the specific role.
- Establish burning platform—Define why the as-is process needs to be changed and prepare a vision that defines how the as-is pain will be lessened by the future-state solution.
- Transformations management—Implement the OCM plan. Test for black holes and lack of acceptance. Train affected personnel in new skills required by the change.

Project Management Approaches

Projects can be managed skillfully or haphazardly. To be a skillful project manager, the individual must acquire the ability to use a large number of approaches effectively. The following is a list of the most common approaches that are in the project management's tool bag (see below).

Additional planning
Additional risk response development
Advertising
Alternative identification
Alternative strategies
Analogous estimating
Benchmarking
Bidder's conferences
Bottom–up estimating
Checklists
Collection
Computerized tools
Conditional diagramming methods
Contingency planning
Contract change control system
Contract negotiation
Contract type selection
Control charts
Cost change control system
Cost estimating tools and techniques
Decision trees
Design of experiments

Earned value analysis
Expected monetary value
Flowcharting
Human resource practices
Independent estimates
Information distribution systems
Information retrieval systems
Insurance
Interviewing
Make-or-buy analysis
Negotiations
Operation definitions
Organization theory
Organizational change management
Parametric modeling
Pareto diagrams
Payment system
Performance reporting tools and techniques
Performance reviews
Preassignment
Procurement
Procurement audits
Quality audits
Quality management plan
Quality planning tools and techniques
Results of quality control measurements
Reward and recognition system
Schedule change control system
Stakeholder analysis
Standard forms
Statistical sampling
Statistical sums
Team-building activities
Training
Trend analysis
Variance analysis
Weighting system
Work around

EXAMPLES

Table 19.1 describes the approaches required to do just one part of managing a project risk management. With a list of approaches so long, it is easy to see that managing a project is not for the weak of heart or the inexperienced (see Tables 19.2 and 19.3).

SOFTWARE

- Project Management.com: $25 per month for one user
- Primavera: $2500 for one user
- Microsoft Project: $995 for one user

TABLE 19.1

Methods Used in Risk Analysis

Method	Description and Usage
Event tree analysis	A hazard identification and frequency analysis technique that employs inductive reasoning to translate different initiating events into possible outcomes.
Fault mode and effects, analysis and fault, and mode effect and criticality analysis	A fundamental hazard identification and frequency analysis technique that analyzes all the fault modes of a given equipment item for their effects both on other components and the system.
Fault tree analysis	A hazard identification and frequency analysis technique that starts with the undesired event and determines all the ways in which it could occur. These are displayed graphically.
Hazard and operability study	A fundamental hazard identification technique that systematically evaluates each part of the system to see how deviations from the design intent can occur and whether they can cause problems.
Human reliability analysis	A frequency analysis technique that deals with the impact of people on system performance and evaluates the influence of human errors on reliability.
Preliminary hazard analysis	A hazard identification and frequency analysis technique that can be used early in the design stage to identify hazards and assess their criticality.
Reliability block diagram	A frequency analysis technique that creates a model of the system and its redundancies to evaluate the overall system reliability.
Category rating	A means of rating risks by the categories in which they fall in order to create prioritized groups of risks.
Checklists	A hazard identification technique that provides a listing of typical hazardous substances or potential accident sources which need to be considered. It can evaluate conformance with codes and standards.
Common mode failure analysis	A method for assessing whether the coincidental failure of a number of different parts or components within a system is possible and its likely overall effect.
Consequence models	The estimation of the impact of an event on people, property, or the environment. Both simplified analytical approaches and complex computer models are available.
Delphi technique	A means of combining expert opinions that may support frequency analysis, consequence modeling, or risk estimation.

(Continued)

TABLE 19.1 (CONTINUED)

Methods Used in Risk Analysis

Method	Description and Usage
Hazard indices	A hazard identification/evaluation technique that can be used to rank different system options and identify the less hazardous options.
Monte Carlo simulation and other simulation techniques	A frequency analysis technique that uses a model of the system to evaluate variations in input conditions and assumptions.
Paired comparisons	A means of estimation and ranking a set of risks by looking at pairs of risks and evaluating just one pair at a time.
Review of historical data	A hazard identification technique that can be used to identify potential problem areas and also provide an input into frequency analysis based on accident and reliability data, etc.
Sneak analysis	A method of identifying latent paths that could cause the occurrence of unforeseen events.

TABLE 19.2

Model Project Plan

Title page

1. Foreword
2. Contents, distribution, and amendment record
3. Introduction
    3.1 General description
    3.2 Scope
    3.3 Project requirement
    3.4 Project security and privacy
4. Project aims and objectives
5. Project policy
6. Project approvals required and authorization limits
7. Project organization
8. Project harmonization
9. Project implementation strategy
    9.1 Project management philosophy
    9.2 Implementation plans
    9.3 System integration
    9.4 Completed project work
10. Acceptance procedure
11. Program management
12. Procurement strategy
13. Contract management
14. Communications management
15. Configuration management
    15.1 Configuration control requirements
    15.2 Configuration management system
16. Financial management
17. Risk management
18. Project resource management
19. Technical management
20. Test and evaluation
21. Reliability management
    21.1 Availability, reliability, and maintainability (ARM)
    21.2 Quality management
22. Health and safety management

TABLE 19.3

Project Processes and Phases

Processes	Conception	Development	Realization	Termination
		Phases		
Strategic Project Process				
Strategic project process	●	X	X	X
Operational Process Groups and Processes within Groups				
Scope-Related Operational Processes				
Concept development	●			
Scope definition	●	X		
Task definition	X	●	X	
Task realization		●	●	X
Change management		●	●	
Time-Related Operational Processes				
Key event schedule planning	X	●	X	
Activity dependency planning	X	●		
Duration estimation	X	●		
Schedule development		●	X	
Schedule control		X	●	X
Cost-Related Operational Processes				
Cost estimation	●	X		
Budgeting		●	X	
Cost control		X	●	X
Resource-Related Operational Processes (Except Personnel)				
Resource planning	X	●		X
Resource control		X	●	X
Personnel-Related Operational Processes				
Organizational structure definition	X	●	●	
Responsibility identification and assignment	X	●	X	
Staff planning and control		X	●	X
Team building	X	●	●	X

(Continued)

TABLE 19.3 (CONTINUED)

Project Processes and Phases

Processes	Phases			
	Conception	Development	Realization	Termination
Communication-Related Operational Processes				
Communication planning	x	●		
Meeting management	x	●	●	x
Information distribution		●	●	x
Communication closure			x	●
Risk-Related Operational Processes				
Risk identification	●	●	x	
Risk assessment	●	●	x	
Solution development		●	x	
Risk control		x	●	
Procurement-Related Operational Processes				
Procurement planning	x	●		
Requirements documentation	x	●		
Supplier evaluation	x	●		
Contracting		●	x	
Contract administration		x	●	x
Product-Related Operational Processes				
Design	x	●		
Procurement	x	●		
Realization			●	
Commissioning				●
Integration-Related Operational Processes				
Project plan development	●			
Project plan execution		●	●	●
Change control		●	●	
Supporting Processes				

Note: ●, key process in the phase; x, applicable process in the phase.

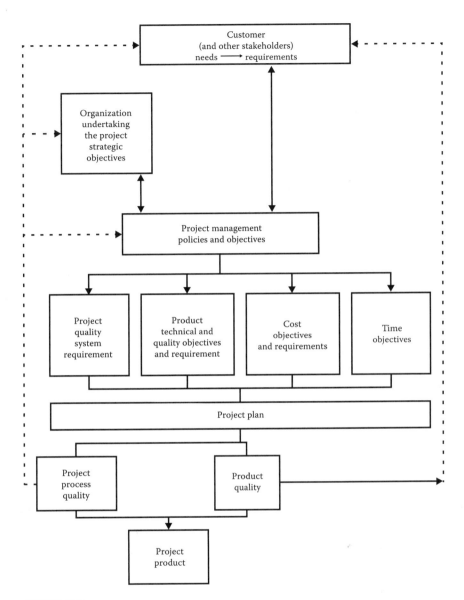

FIGURE 19.3
Block diagram of the project management process.

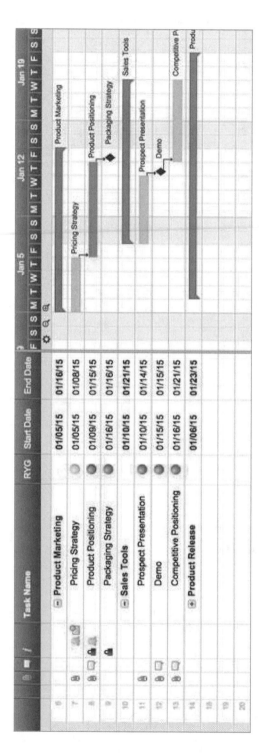

FIGURE 19.4

Gantt chart project timeline example.

SUGGESTED ADDITIONAL READING

Badiru, A.B. and Whitehouse, G.E. *Computer Tools, Models, & Techniques for Project Management.* Blue Ridge Summit, PA: TAB Books, 1989.

Block, R. *The Politics of Projects.* New York: Yourdon Press, 1983.

Brooks, H.E., Jr. *Project Management in the Information Technology Age.* Downey, CA: Sterling Series, 1989.

Cleland, D. and Roland G., eds. *Global Project Management Handbook.* New York: McGraw-Hill, 1994.

Dinsmore, P. *Human Factors in Project Management.* New York: AMACOM Publications, 1990.

Dinsmore, P.C. *The AMA Handbook of Project Management.* New York: AMACOM Books, 1993.

Einsiedel, A. *Improving Project Management.* Englewood Cliffs, NJ: Prentice-Hall, 1988.

Focardi, S. and Jonas, C. *Risk Management: Framework, Methods and Practice.* New York: McGraw-Hill, 1998.

Harrington, H.J. and McNellis, T. *Project Management Excellence.* Chico, CA: Paton Press LLC, 2006.

Harrington, H.J., Conner, D.R., and Horney, N. *Project Change Management.* New York: McGraw-Hill, 1999.

ISO/DIS 10006: Guidelines to Quality in Project Management. Geneva, Switzerland: International Organization for Standardization, 1998.

MatchWare.com. *Project Management: Fast and Effective Project Planning.* Retrieved June 10, 2013.

PMI Standards Committee. *A Guide to the Project Management Body of Knowledge.* Upper Darby, PA: Project Management Institute, 1996.

Project Management Institute. *A Guide to the Project Management Body of Knowledge.* Sylva, NC: Project Management Institute, 1996.

Project Management Institute, *A Guide to the Project Management Body of Knowledge: PMBOK(R) Guide.* Paperback Project Management Institute, January 1, 2013.

Westney, R.E. *Computerized Management of Multiple Small Projects.* New York: Marcel Dekker Inc., 1992.

20

S-Curve Model

Achmad Rundi and Frank Voehl

CONTENTS

DEFINITION

The S-curve is a mathematical model also known as the logistic curve, which describes the growth of one variable in terms of another variable over time. S-curves are found in many fields of innovation, from biology and physics to business and technology.

USER

This tool can be used by individuals but its best use is with a group of four to six people. Cross-functional teams usually yield the best results from this activity.

OFTEN USED IN THE FOLLOWING PHASES OF THE INNOVATIVE PROCESS

The following are the seven phases of the innovative cycle. An X after the phase name indicates that the tool/methodology is used during that specific phase.

- Creation phase X
- Value proposition phase X
- Resourcing phase
- Documentation phase
- Production phase
- Sales/delivery phase X
- Performance analysis phase X

TOOL ACTIVITY BY PHASE

- Creation and value proposition phases—The S-curve is used to predict when a new creative cycle needs to be implemented. As the S-curve starts to drop off, it signals the need for a new product line and the start of the creativity phase.
- Sales/delivery phase—The S-curve is used during this phase to determine when the market is saturated, and the revisions or upgrades to the product that need to be implemented to maintain an acceptable level of sales. The slope of the S-curve is also used to determine when a new product line needs to start the creativity phase.
- Performance analysis phase—Technology businesses, such as computer, software, and electronic manufacturers, often display an S-curve performance analysis life cycle. That initial progress is slow because the principles of the technology are poorly understood. Once researchers get a better understanding of the technology capabilities, progress accelerates rapidly.

HOW TO USE THE TOOL

Introduction

The S-curve is a mathematical model also known as the logistic curve, which describes the growth of one variable in terms of another variable over time. S-curves are found in many fields of innovation, from biology and physics to business and technology. In terms of innovation in business, the S-curve is used to describe—and sometimes predict—the performance of a company or a product over a period of time. As such, the S-curve portrays the general pattern in which a successful organization starts out with a new product with only a small initial group of customers. This is followed by a period of rapid growth as the general public starts to seek out the new product, and eventually the demand for the product peeks out and levels off as the market matures and the population of new customers declines. As the demand for the product is fulfilled, sales drop and growth is slow and sometimes negligible. Demand is kept from collapsing by offering new features or reducing prices to maintain a share of the customer base. When the S-curve starts to drop off, it is time for the organization to introduce a replacement product or service.

The concept of the S-curve was introduced in the business in the late 1880s. As the result of the work done by Stanford University professor Everett Rogers and the book he published entitled *Diffusion of Innovation*, it became a standard in most businesses and university business and engineering courses in the 1960s. Rogers' work showed how the accumulated sums of the customer buying patterns resembled the shape of the letter S. His terminology is still currently used today and is reflected in many business books and college curriculums (see Figure 20.1).

Figure 20.1 depicts the rate of adoption in terms of the five stages of adoption: innovators, early adopters, early majority, late majority, and laggards.

Using the Tool

1. *Using the tool in innovation project management.* S-curves are an important project management tool in that they allow the progress of a project to be tracked visually over time, and form a historical record of what has happened to date. Analyses of S-curves allow

S-curve and innovation

- S-curve is a measure of the speed of adoption of an innovation.

- First used in 1903 by Gabriel Tarde, who first plotted the S-shaped diffusion curve.

- This process that has been proposed as the standard life cycle of innovations can be described using the "S-curve."

Adoption and the S-curve

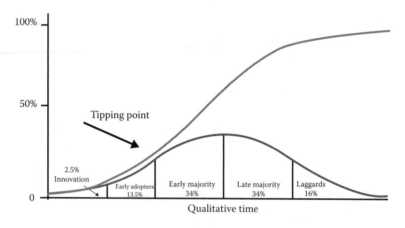

FIGURE 20.1
Adoption and the S-curve.

project managers to quickly identify project growth, slippage, and potential problems that could adversely affect the project if no remedial action is taken.

Comparison of the baseline and target S-curves quickly reveals if the project has grown (the target S-curve finishes above the baseline S-curve) or contracted (the target S-curve finishes below the baseline S-curve) in scope. A change in the project's scope implies a reallocation of resources (increase or decrease), and the very possible requirement to raise contract variations. If the resources are fixed, then the duration of the project will increase (finish later) or decrease (finish earlier), possibly leading to the need to submit an extension of time claim (see Figure 20.2).

2. *Determining innovation development slippage.* This is defined as, "The amount of time a task has been delayed from its original baseline schedule. The slippage is the difference between the scheduled

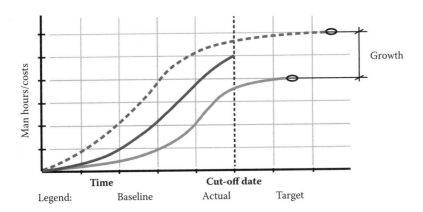

FIGURE 20.2
Calculating project growth using S-curves.

start or finish date for a task and the baseline start or finish date. Slippage can occur when a baseline plan is set and the actual dates subsequently entered for tasks are later than the baseline dates or the actual durations are longer than the baseline schedule durations."

Comparison of the baseline S-curve and target S-curve quickly reveals any project slippage (i.e., the target S-curve finishes to the right of the baseline S-curve). Additional resources will need to be allocated or additional hours worked in order to eliminate (or at least reduce) the slippage. An extension of time claim may need to be submitted if the slippage cannot be eliminated or reduced to an acceptable level (see Figure 20.3).

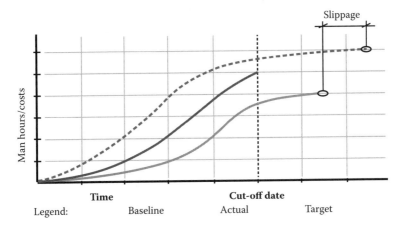

FIGURE 20.3
Calculating project slippage using S-curves.

3. *Determining progress by using a comparison of the target S-curve and actual S-curve to reveal the progress of the project over time.* In most cases, the actual S-curve will sit below the target S-curve for the majority of the project (due to many factors, including delays in updating the production schedule). Only toward the end of the project will the curves converge and finally meet. The actual S-curve can never finish above the target S-curve. If the actual S-curve sits above the target S-curve at the cut-off date, the production schedule should be examined to determine if the project is truly ahead of schedule, or if the production schedule contains unrealistic percentage complete values for ongoing tasks (see Figure 20.4).

4. *Using the S-curve to portray the general pattern between a new product and customer demand.* Most successful organizations start out with a new product with only a small initial group of customers. This is followed by a period of rapid growth as the general public starts to seek out the new product. Eventually the demand for the product peeks out and levels off as the market matures. As the demand for the product is fulfilled, sales drop off. To offset this downward trend, organizations frequently start offering new features or reduce prices to the customer base. When the S-curve starts to drop off, it is time for the organization to introduce a replacement product/service.

Generally, it is an S-shaped curve that shows the growth of a variable against another variable, usually in the unit of time. The S-curve

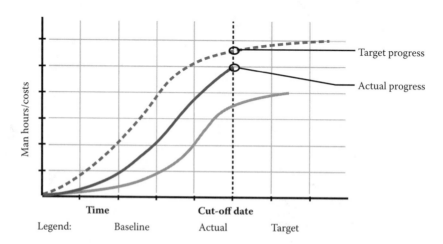

FIGURE 20.4
Calculating project progress using S-curves.

would start at low increase, then continue with spike/exponential increase in growth, and then follow by tapering growth. The beginning represents a slow, deliberate but accelerating start, while the end represents a deceleration as the variable runs out room for growth (Tawfeq, 2011).

In innovation, the S-curve can show the growth of performance and the rate of diffusion (Schilling, 2010). The performance S-curve shows the growth of performance of the innovation, whether it be the product or the technology behind this innovation against variable of time or combination of time and cost.

The slow but accelerating initial growth is due to the limited amount of investment and research and development spend on the innovation. Then, as more research and development (R&D) and investment are put into this innovation, it takes off exponentially at a much faster rate. As there are no more ways to improve the performance of the product or the technology behind the innovation, the growth decelerate, and cost and time spent will not be added value as before, as diffusion of innovation at Apple shows (Figure 20.5; Eriksson, 2013).

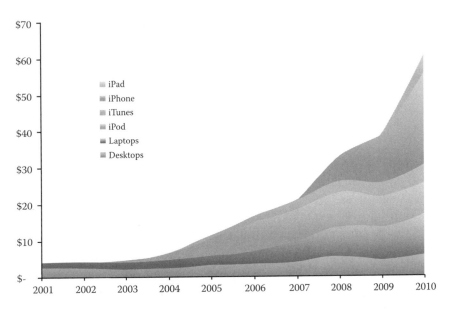

FIGURE 20.5
Innovation S-curve at Apple during a 10-year period. (From the 2012 LIFT Conference. See http://www.slideshare.net/wright4/double-scurve-model-of-growth.)

5. *The innovation diffusion S-curve shows the rate of adoption vs. disruption over time in term of speed against time* (Schilling, 2010). This S-curve has the same phases to performance S-curve but it shows how many adopted in early period due to its newness, and exponential growth of amount in adoption due to awareness and consumers' readiness toward the innovation, and tapering growth due to other new alternatives or innovation. The type of the adopters can be explained in a separate bell curve, as established by Everett Rogers in terms of early adopters to laggards. The performance and diffusion S-curve is interrelated but not the same. It is interrelated to show that the more interests and enough market insight there is, there is more investments toward R&D that will improve the innovation's performance (Schilling, 2010).

The S-curve is cyclical. When one S-curve tapered off, there is another S-curve that starts its initial phase.

- Get used to map the progress of the innovation performance, plotting the performance/adoption rate level against time and investment.
  - Data can be on own performance and investment of own innovation, or data of average performance of multiple producers in the industry and industry's investments (Schilling, 2010).
- As we track for early spike and tapering, we will know when to jump. As we jump, there are prerequisites that we need to prepare or consider. Before then, we also need to find new innovation ideas based on big enough market insights (Nunes and Breene, 2011) so that we can develop the new innovation when the time is right.

The S-curve is cyclical. When one S-curve tapered off, there is another S-curve that starts its initial phase (see Figure 20.6).

A company can be successful, as shown by the trail of its S-curve, but the high achiever company will be far-sighted and will envision the next new S-curves that come after each of the others' fade, as illustrated by the Apple 10-year period in Figure 20.5 and Sony in Figure 20.6. These companies will have the strategy to jump the S-curve early on before the sign of tapering begins to manifest itself in disruptive innovation, as shown in Figure 20.7.

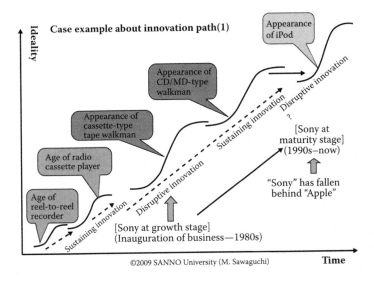

FIGURE 20.6

S-curve cycles. (From Nakagawa, T. *TRIZ Forum: Conference Report (22-E)*. Retrieved from TRIZ Forum. December 13, 2009.)

Disruptive innovation

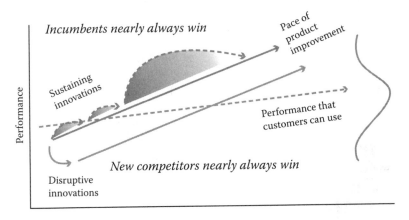

FIGURE 20.7

Disruptive innovation S-curve. (From Christensen, C. M. and Raynor, M. E. *The Innovator's Solution: Creating and Sustaining Successful Growth.* Massachusetts: Harvard Business School of Publishing Corporation, 2013.)

Illustrations

In the innovation management field, the S-curve illustrates the intro-duction, growth, and maturation of innovations as well as the techno-logical cycles that most industries experience. In the early stages, large amounts of money, effort, and other resources are expended on the new technology, but small performance improvements are observed. Then, as the knowledge about the technology accumulates, progress becomes more rapid. As soon as major technical obstacles are overcome and the innovation reaches a certain adoption level, an exponential growth will take place, as Table 20.1 illustrates in the case of the disruptive innova-tion S-curve.

Table 20.1 illustrates the impact of the disruptive innovation S-curve in the case of four innovations and markets shown. During this phase, relatively small increments of effort and resources will result in large per-formance gains. Finally, as the technology starts to approach its physical limit, further pushing the performance becomes increasingly difficult, as Figure 20.8 shows.

Consider the supercomputer industry, where the traditional architec-ture involved single microprocessors. In the early stages of this technol-ogy, a huge amount of money was spent in research and development, and it required several years to produce the first commercial prototype. Once the technology reached a certain level of development, the know-how and expertise behind supercomputers started to spread, boosting dramati-cally the speed at which those systems evolved. After some time, however, microprocessors started to yield lower and lower performance gains for a given time/effort span, suggesting that the technology was close to its physical limit (based on the ability to squeeze transistors in the silicon

TABLE 20.1

Disruptive Innovation S-Curve

Innovation	Disrupted Market	Details of Disruption
Digital photography	Chemical-based photography	First examples of digital cameras were of very poor quality and laughed at
LCD	CRT	LCDs were first monochromatic and of low resolution
Wikipedia	Traditional encyclopedias	*Encyclopaedia Britannica* ended print production in 2012
LED lights	Light bulbs	Initial LED lights were only strong enough to be indicators, but now replacing most lighting

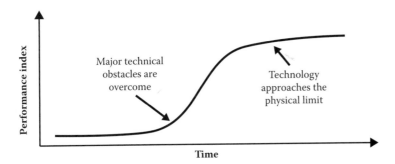

FIGURE 20.8
Innovation technology management S-curve.

wafer). To solve the problem, supercomputer producers adopted a new architecture composed of many microprocessors working in parallel. This innovation created a new S-curve, shifted to the right of the original one, with a higher performance limit (based instead on the capacity to coordinate the work of the single processors) (see Figure 20.9).

Usually the S-curve is represented as the variation of performance in function of the time/effort. Probably, that is the most used metric because it is also the easiest to collect data for. This fact does not imply, however, that performance is more accurate than the other possible metrics, for instance the number of inventions, the level of the overall research, or the profitability associated with the innovation.

One must be careful with the fact that different performance parameters tend to be used over different phases of the innovation. As a result, the outcomes may get mixed together, or one parameter will end up

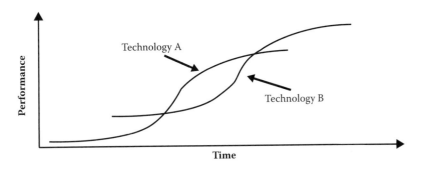

FIGURE 20.9
Supercomputer industry S-curve.

influencing the outcome of another. Civil aircraft provides a good example. In early stages of the industry, fuel burn was a negligible parameter, and all the emphasis was on the speed aircrafts could achieve and if they would thus be able to get off the ground safely. Over time, with the improvement of the aircrafts, almost everyone was able to reach the minimum speed and to take off, which made fuel burn the main parameter for assessing the performance of civil aircrafts. Overall, we can say that the S-curve is a robust yet flexible framework to analyze the introduction, growth, and maturation of innovations, and to understand the technological cycles. The model also has plenty of empirical evidence; it was exhaustively studied within many industries, including semiconductors, telecommunications, hard drives, photocopiers, jet engines, and so on. Figure 20.10 illustrates the process of what it takes to climb and jump the S-curve in business innovation.

Figure 20.10 demonstrates the process for climbing and jumping the S-curve, and is based on the information found in the Accenture whitepaper at http://assets-production.govstore.service.gov.uk/G4/Accenture _UK_Limited-0360/5236e49b3540676c59f07b50/QD1/Accenture%20 Cloud%20Project%20and%20Programme%20Management%20-%20 Service%20Definition%20Document.pdf.

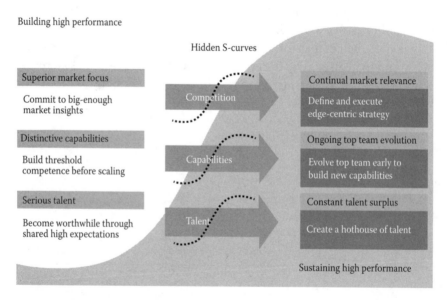

FIGURE 20.10
Model for climbing and jumping the S-curve.

SUMMARY

The use of the term S-curve in business circles goes back at least to the late 1800s, but it gained broader currency in the early 1960s, when Stanford University professor Everett Rogers published *Diffusion of Innovations*. In this book, Rogers shows how the cumulative sum of adopters of an innovation takes on the shape of the letter S. He then characterized segments of adopters in terms that are still commonly used today.

During the dot-com era, the meaning of the phenomenon was adapted to describe the rollout of new Internet-related technologies. Best-selling books like Geoffrey Moore's *Crossing* the Chasm explained how new-technology companies could break out of the bottom of the curve and reach the top. Accenture* and other consultants have found that the term S-curve can be adapted yet again to explain business performance over time. Companies thrive, after all, by successfully delivering some form of innovation to customers. Performance starts slowly as the business is launched, and the company experiments to find the right business formula. Then performance accelerates rapidly as word of the attractiveness of the offering spreads, and finally it fades as the market approaches saturation, imitators appear, and obsolescence leads to better substitutes.

EXAMPLE

The following is a S-curve case study. Cray Inc. has been in the supercomputer industry for many years, and it depended earlier on traditional architecture involving a single microprocessor. It went through the early phase with turbulence and trials, where huge amounts of money were invested during R&D. It spent several years before it could produce the first commercial prototype (Eriksson, 2013).

When Cray Inc. was able to reach a certain development level, expertise and knowledge behind supercomputer began to spread. This allowed the speed of the of the supercomputer system to improve exponentially. Eventually, the performance of the microprocessor technology yielded

---

* See the Accenture classic brief, "Jumping the S-curve: How to beat the growth cycle, get on top, and stay there," by Paul Nunes and Tim Breene (2012).

lower and lower gain for a given investment against time span. This showed that the technology was nearing its limit.

Along the time, Cray Inc. had reviewed and adopted an innovative way to solve the microprocessor issue and was able to apply architecture that allowed multiple processors working in parallel. This effort allowed Cray Inc. to have a new S-curve, a shift to right in the S-curve diagram from the original S-curve, with a higher rate of performance.

SOFTWARE

- S-Curve for Microsoft Project is used to plot S-curves for a single, master project or selected activities. Three curves are presented, baseline, actual, and schedule for a given MSProject. S-Curve can help you track your MSProjects visually using an S-curve graph. Progress and schedule are shown against base. For details, see http://download.cnet .com/SCurve-for-Microsoft-Project/3000-2076_4-10545060.html.

- Promineo is the developer of the S-curves for Primavera Software application and is authorized to license the S-curves for Primavera Software. Subject to the user's compliance with this end user license agreement (EULA), Promineo grants the user a nonexclusive, nontransferable, limited, revocable license to use the software solely according to this EULA. The license entitles the user to use a limited number of copies of the software; the number is defined in the registration information, or in a purchase order. Promineo reserves all rights not expressly granted to the registered user. Promineo retains the ownership of the software. The software is protected by copyright laws and international copyright treaties, as well as other intellectual property laws and treaties. For details, see http://www.primaveraworks.com/?gclid=Cj0KEQiAla e1BRCU2qaz2_t9IIBEiQAKRGDVfomAZxOyl7ZWNNg8iafJzrVg8Z -AlB5J75Z47ovmdUaAiom8P8HAQ.

REFERENCES

Christensen, C.M. and Raynor, M.E. *The Innovator's Solution: Creating and Sustaining Successful Growth.* Massachusetts: Harvard Business School of Publishing Corporation, 2013.

Eriksson, J. *The S-curve for Apple is flattening*. Retrieved from Bearing Consulting Blog: http://blog.bearing-consulting.com/2013/01/25/the-s-curve-for-apple-is-flattening/, January 25, 2013.

Nakagawa, T. TRIZ Forum: Conference Report (22-E). Retrieved from TRIZ Forum, December 13, 2009.

Nunes, P. and Breene, T. *Jumping the S-Curve*. Retrieved from Accenture: http://www.accenture.com/sitecollectiondocuments/pdf/accenture_jumping_the_s_curve.pdf, 2011.

Schilling, M.A. *Strategic Management of Technological Innovation*. New York: McGraw-Hill, 2010.

Tawfeq, M. *What is the S-Curve, and how do calculate the work progress*. Retrieved from Planning Planet: http://www.planningplanet.com/forums/planning-scheduling-programming-discussion/502846/what-s-curve-and-how-do-calculate-work-prog, March 28, 2011.

SUGGESTED ADDITIONAL READING

6–3–5 Brainwriting. Retrieved from Wikipedia.com: http://en.wikipedia.org/wiki/6-3-5_Brainwriting, n.d.

635–369. Retrieved from Nagoya University of Art and Science: http://media.nuas.ac.jp/~robin/zakka/z-lesson/369.htm, n.d.

Beard, R. Competitor analysis template: 12 Ways to predict your competitors' behaviors. Retrieved from *Client Heartbeat*: http://blog.clientheartbeat.com/competitor-analysis-template/, May 3, 2013.

Besanko, D., Dranove, D., Schaefer, M., and Shanley, M. *Economics of Strategy*. New York: John Wiley & Sons, 2010.

Czeipiel, J.A. and Kerin, R.A. Competitor analysis. In *Handbook of Marketing Strategy*, Shankar, V. and Carpenter, G.S. (eds.). Northampton, MA: Edward Elgar Publishing, 2011.

Frost, A. Knowledge management processes. Retrieved from KMT: An Educational KM Site: http://www.knowledge-management-tools.net/knowledge-management-processes.html, 2010.

Harrington, H.J. *Knowledge Management Excellence: The Art of Excelling in Knowledge Management*. Chico, CA: Paton Professional, 2011.

Koenig, M.E. What is KM? Knowledge management explained. Retrieved from KMWorld: http://www.kmworld.com/Articles/Editorial/What-Is-.../What-is-KM-Knowledge-Management-Explained-82405.aspx, May 4, 2012.

linkmv97. The 6–3–5 method (BrainWriting). Retrieved from Youtube.com: http://www.youtube.com/watch?v=p-I6a6AqDBM, April 28, 2009.

McAdam, R. Knowledge management as a catalyst for innovation within organizations: A qualitative study. *Knowledge and Process Management*, pp. 233–241, 2000.

Nonaka, I. and Toyoma, R. A Firm as a Dialectical Being: Towards a Dynamic Theory of a Firm. *Industrial and Corporate Change*, vol. 11, pp. 995–1009, n.d.

Power, D. *The Curve Ahead: Discovering the Path to Unlimited Growth*. New York: St. Martin's, 2014.

Stacey. 08 Ideas in 30 minutes—The 6–3–5 method of brainwriting. Retrieved from Blog-session: http://blogsession.co.uk/2014/03/635-method-brainwriting/, March 25, 2014.

Wilson, C. Using brainwriting for rapid idea generation. Retrieved from Smashing Magazine.com: http://www.smashingmagazine.com/2013/12/16/using-brainwriting -for-rapid-idea-generation/, December 16, 2013.

21

Safeguarding Intellectual Property

Steven G. Parmelee

CONTENTS

Abraham Lincoln got it exactly right when he said that the patent system "added the fuel of interest to the fire of genius in the discovery and production of new and useful things."

DEFINITION

The expression *intellectual property rights* refers to a number of legal rights that serve to protect various products of the intellect (i.e., *innovations*). These rights, while different from one another, can and do sometimes offer overlapping legal protection.

USER

This tool can be used by individuals, groups, and any of a variety of legal enterprises (partnerships, limited liability corporations, public corporations, nonprofit organizations, etc.). These concepts are also usable by either those who innovate or those who hire the innovators.

OFTEN USED IN THE FOLLOWING PHASES OF THE INNOVATIVE PROCESS

The following are the seven phases of the innovative cycle. An X after the phase name indicates that the tool/methodology is used during that specific phase.

- Creation phase X
- Value proposition phase X
- Resourcing phase
- Documentation phase
- Production phase X
- Sales/delivery phase X
- Performance analysis phase X

TOOL ACTIVITY BY PHASE

Timing can be everything when it comes to perfecting an intellectual property right or at least beginning the process, and the time all too frequently is *now*. In many countries (including the United States), for example, when multiple parties seek a patent on the same basic idea, it is the first party who files their patent application who will typically win the right to receive the patent. So whether it be an interest in patents, trademarks, trade secrets, or copyrights, the identification and perfection process is often an ongoing one that begins with the glimmer of an idea and that often extends well into the more mature phases of the innovation.

HOW TO USE THE TOOL AND EXAMPLES

The expression *intellectual property rights* refers to a number of legal rights that serve to protect various products of the intellect. Typical examples of intellectual property rights include patents, copyrights, trade secrets, and trademarks. As a result, and unfortunately for the innovator and entrepreneur, it can become confusing and a challenge to understand whether and when a particular intellectual property right applies with respect to a given innovation.

Each intellectual property right represents a complex and nuanced area of legal practice. It would be impossible for a short chapter such as this to comprehensively treat even a single intellectual property right. There are, however, a number of general points and practices that every innovator should understand (and ignore at their own peril). This chapter therefore aims to begin building the reader's initial introduction and understanding of a few of the more common intellectual property rights.

There are many ways to explore this topic. And, unfortunately, many of those ways tend to combine a seemingly dry subject matter with an overly turgid style of presentation. It is often said that we learn best from our own mistakes. It has also been this author's experience that many folks can learn nearly as well from the mistakes of others. With the latter in mind, many concepts and practice tips (i.e., *tools*) pertaining to various intellectual property rights will be brought forth in this chapter in the context of a simple vignette, in particular through the story of Pauline Pastures and her amazing cookies.

First, however, a bit of housekeeping. Nothing stated in this chapter is offered as legal advice. The author reserves the right to have opinions different than those presented here given a specific fact situation, changes in the law, changes in the practice of law, or simply due to a change of mind. Readers considering acting upon their own opportunities to acquire intellectual property rights should seek competent legal counsel.

And now, on with the show!

Idea (Pauline's Case Study)

Pauline Pastures is a homemaker. One afternoon, while baking cookies, she spilled some flour on the floor. While her cookies baked in the oven, she used her vacuum cleaner to clean her kitchen. A short while later, when sampling the cookies that had been baking at the time of the foregoing activities, she noticed that the cookies were incredibly delicious! Pauline was an experienced cook and had baked many a batch of cookies. So it is not just that the cookies were good, but that they were uniquely better than any other cookies she had ever made.

A cookie this good was worth investigating. She experimented with further batches while trying to isolate or otherwise identify what, exactly, had led to this excellent result. Curiously, during the next few days, she discovered that she could only obtain this delicious result when using her vacuum cleaner during the baking process. During the course of this experimentation, she also discerned that her other *delicious batches*, while indeed delicious, differed from one another with respect to just how delicious they were.

TIME-OUT!

One intellectual property right concerns patents. Patents are a grant from a national government or a regional government entity (such as the European Patent Office) that provides the patentee (i.e., the person or company who owns the patent) with the right to prevent others from making, using, or selling the patented invention. An invention must be reduced to practice in order to qualify for a patent. While this does not necessarily mean that the inventor must have built the invention, it typically means that the inventor must be able to explain the invention in a patent application to a degree sufficient to ensure that readers of the patent could themselves make and practice the invention. At this point in Pauline's story, she is possibly now on the ragged edge of being able to meet that standard.

Pauline continued to experiment in the coming weeks and months. She discovered, through trial and error, the optimum times and ways to use the vacuum cleaner to obtain, consistently, the best-tasting cookies. As illustrated in Figure 21.1, Pauline's optimum *recipe* specifies that halfway through the total baking time she run the vacuum cleaner within three to five feet of the baking cookies for a period of time equal to one-fourth the total baking time (see Figure 21.1). Closer or farther, longer or shorter, and the desired results are not obtained.

Pauline, in fact, realized that her discovery might be worth something. It seemed at least a reasonable possibility that people might be willing to pay good money for cookies that tasted this delicious.

So she decided to protect her idea by writing it down, putting that writing into a self-addressed envelope, and mailing it to herself. She had read or heard about this *poor man's patent*, and how the cancellation date on the stamp could serve to prove that she was the inventor.

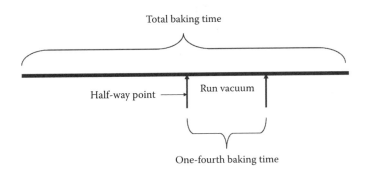

FIGURE 21.1
Results of experimentation related to the baking of cookies.

TIME-OUT!

It is now considerably more likely that Pauline has reduced her invention to practice. All other things being equal, now might be a good time for Pauline to make a knowing decision about whether to seek patent protection for her process.

A couple of other points are worth mentioning here. First, Pauline does not understand why her process works, only that it does apparently work. Second, she did not achieve her results through great brilliance, education, or the like. Instead, she made a mistake, noticed something by coincidence, and conducted a lot of experiments to figure out how to replicate that something. And that is all okay! There is no requirement for most if not all patent systems that the inventor understands why the invention works, so long as the inventor is able to explain how to make and use the invention. There also is no requirement that the inventor *invent* his or her invention via any particular process. (In the United States many decades ago, the Supreme Court once ruled that an invention must be the result of an inventive flash of brilliance; the U.S. Congress modified the patent statutes to do away with such a requirement.) Accordingly—and this is important—Pauline is an *inventor*.

TIME-OUT!

Sorry Pauline, but this practice did little or nothing in the past and certainly offers no particular utility today. In the past, when U.S. patent law protected the first person to invent a given invention, this so-called poor man's patent carried little or no weight in court when determining ownership or dates of inventorship. It was simply too easy to cheat! Consider, for example, that an *unsealed*, self-addressed envelope could be sent to oneself from time to time, and then later stuffed and sealed with whatever this person might choose to insert. As for present times, the United States recently switched from a first-to-invent system to a first-to-file system, where *file* refers to filing an appropriate application at the U.S. Patent and Trademark Office. Under this first-to-file system, the old poor man's patent has lost even further ground and simply has no place in any realistic effort to protect one's inventions.

On the other hand, Pauline's writing is a copyrightable work of authorship. This is because essentially any original work of authorship (or work of art) is protectable by copyright. Even better, her writing is already protected by copyright even if she fails to include a copyright notice on it. This is because copyright arises when a protectable work of authorship is first rendered in tangible form, and that right will last for decades!

So, just what does Pauline's copyright get her? Copyright law gives Pauline the right to control whether and how other people make or use a copy of her written recipe. This means she can control the making of copies of that document, the making of derivative versions of that copy (such as a foreign language translation of her recipe, or a revised recipe that doubles all of the quantities to increase the size of the resultant batch), public performances to the extent that such a thing is possible, and so forth.

In some countries, such as the United States, Pauline could register her claim to copyright in this written document. In the United States, this entails filling out a government form and providing that form to the Library of Congress along with copies of the written recipe itself and a nominal filing fee. Early registration, in turn, is the predicate for some potentially significant benefits. For example, early registration can help preserve the right to seek so-called statutory damages against an infringer. Statutory damages represent potentially high monetary damages that often shock the alleged infringer (statutory damages being the typical way, for example, for the recording industry to demand millions of dollars from college students who unlawfully downloaded only a few sound recordings from the Internet). Early registration can also establish a basis for the copyright owner to demand attorney's fees from an infringer. Anyone who has dealt much with attorneys will likely quickly understand that being held accountable for the other side's attorney's fees might be more financially debilitating than the other damages at issue.

Cheap, fast, and easy, copyright sounds great! So, there must be a catch, right?

If there is a catch, perhaps it is this: *Copyright only protects the expression of an idea and not the idea itself.* As a result, a person who reads Pauline's written recipe might well be free and clear to begin baking cookies using Pauline's described process, as baking the cookies might be viewed as constituting the idea while the recipe constitutes the expression of that idea. Other people might also be free to read Pauline's recipe and then reexpress that recipe using their own words in, say, a cookbook. Generally speaking, that may be fair game.

Marketing Plan

Pauline decides that she has little interest in running her own cookie business. She therefore writes a letter to the Titan Cookie Company (she has always loved the little animated magic toadstool cookie bakers they feature in their television advertisements) and describes her recipe. She explains that she would be willing to accept only a fairly modest royalty in exchange for their right to make these cookies. Her letter concludes by asking them how much they would be willing to pay to use her recipe and expressing hopes for a mutually beneficial future together.

TIME-OUT!

Oh, dear.

There is a decent chance that Pauline has now probably messed up her trade secret rights to her process. Unlike patents and copyrights, trade secrets are more typically protected (in the United States) via individual states (sometimes by statute and sometimes by common law as enforced by the courts). Generally speaking, the courts will protect the confidentiality of information having some competitive value so long as the owner takes reasonable precautions to maintain that confidentiality. Unfortunately, in many cases, once this particular horse is out of the barn, it is very hard to put it back.

Trade secrets are protected by keeping the secret a secret. What might surprise some readers is that a secret is not defined by how few, or how many, people know the secret. Instead, a trade secret can be protected by only sharing the secret with others who are bound to maintain the secret. It is possible to compromise a trade secret by sharing that secret with only a single person who is not obligated to maintain the secret. Similarly, a trade secret may safely remain a protectable secret even if thousands of people know the secret so long as every one of those individuals is legally obligated to maintain the secret.

A nondisclosure agreement (often called an NDA in the industry) is a legal contract that typically primarily serves to preserve one or more trade secrets that one party is going to share with another. Such agreements often identify (either generally or specifically) the secret subject matter, the duration of time that the secret must be maintained, and any restrictions regarding any use that the party receiving the secret may be able to enjoy while maintaining the secret. Depending on the circumstances, these agreements sometimes include many other terms and conditions as well.

The courts will also sometimes enforce a trade secret in a situation where a confidential relationship leading to a reasonable expectation of confidentiality can at least be implied. Relying on such a happenstance, however, is often a poor up-front strategy, and specific agreements to maintain a secret are typically far better suited to the task.

Trade secrets can be a very important part of an overall intellectual property strategy and a very cost-effective tool at that. Businesses buy and sell confidential information all the time. Once that secret becomes public, however, protection as a trade secret usually disappears along with the cloak of secrecy.

Paula's letter may also have caused some interesting things to happen in patent land. Perhaps foremost among these, a disclosure of an invention by an inventor effectively begins a one-year clock that counts down a grace period during which the inventor can file a U.S. patent application before essentially losing the right to file that application. As a result, whether she knows it or not, Pauline now likely has one year to file a U.S. patent application or she may very well lose that opportunity.

Disappointment

The Titan Cookie Company responds to Pauline's letter with a letter of its own.

First, they include with their letter Pauline's written description (the original, not a copy).

Second, they explained that they have a policy of not reviewing or considering outside idea submissions without the submitting party having signed a nonconfidentiality agreement.

Third, they provide a blank nonconfidentiality agreement for Pauline to employ along with instructions on how to resubmit her idea.

TIME-OUT!

In at least one regard, Pauline is fortunate. The Titan Cookie Company is behaving like a typical large U.S. company and hence is unlikely to steal her submitted idea (at least, if you are particularly cynical, in this particular way and at this particular time). It is very possible that her letter never made it past a clerk, paralegal, or company attorney whose job is to intercept and return such unsolicited inputs as described above. These practices are designed to protect the company from allegations of trade secret theft that are based, for example, on an implied confidential relationship.

Nonconfidential disclosure agreements create a relationship that explicitly denies the existence of a confidential setting. In short, trade secret rights, if any, are not *secrets* anymore and hence likely unenforceable at law. Just as a nondisclosure agreement makes certain (at least to the extent of the clarity and sufficiency of its own terms and conditions) the existence of a trade secret, a properly drafted nonconfidential disclosure agreement will accomplish exactly the opposite. This does not mean that an innovator should never sign such an agreement. It is more than appropriate, however, that the innovator who signs such an agreement be fully aware and cognizant of what it is they are surrendering.

Pauline's Time-Out

This response from the Titan Cookie Company discourages Pauline. She puts the letter aside, along with that unsigned nonconfidential disclosure agreement, and goes on with life for a while without thinking much of her cookie process.

Marketing Plan—Take Two!

Eventually, though, about a year later, Pauline's entrepreneurial spirit reasserts itself. Pauline begins to sell her cookies through a Yahoo website storefront that her nephew set up for her (in exchange, of course, for a batch of Pauline's cookies). Things start slowly and sales remain quite manageable; Pauline keeps up with her orders by baking her own cookies.

Pauline also continues to experiment at home with her process. At one point, she decides to try different brands of vacuum cleaners. For reasons she certainly does not understand, cookies baked under the influence of her Remora-brand vacuum cleaner yield cookies that are twice as delicious (as best as she can discern) as the next best brand of vacuum cleaner.

At about this same time, the website www.incrediblecookies.com rates Pauline's cookies as *freaking awesome* and gives her storefront its highest rating—five bites! Pauline's sales skyrocket from this moment on.

To keep up with this increased demand, Pauline hires a local bakery to bake her cookies for her. This means, of course, that Pauline teaches the folks at this local bakery about her vacuum cleaner–based baking process.

TIME-OUT!

Should Pauline perhaps have considered requiring the folks at the bakery to sign a nondisclosure agreement to require them to keep her baking process a secret?

Oh No!

Things, in fact, are moving along just fine. That is, until a worker at the bakery moves on to a new job … at the Titan Cookie Company.

The Titan Cookie Company, for its part, has noticed Pauline's success and high ratings from the incrediblecookies.com website, and has been trying to figure out how she does it. One thing leads to another, and the transplanted worker from Pauline's bakery is soon a rising star at Titan as vacuum cleaners are quickly added to Titan's high-capacity, large-batch cookie baking lines.

The Titan Cookie Company soon introduces its new Xtreme cookie line. These cookies are indeed good, but just as important, Titan's marketing power proves devastating. Pauline's sales plummet.

Finally!

Pauline finds her way to her friendly neighborhood intellectual property attorney, and she relates all of the above details. The attorney listens attentively, takes plenty of notes, and then gives Pauline the bad news about

- Pauline's lost trade secret
- How Pauline's copyright would not solve this problem
- How Pauline has likely lost the right to seek a patent for her basic process because too much time has now passed

TIME-OUT!

Are you starting to see why some people really do not like lawyers very much?

But Pauline's attorney also explains to Pauline that she still has time to seek a U.S. patent for her latest vacuum cleaner improvement (this being the use of the Remora-brand vacuum cleaner).

TIME-OUT!

Copyrights are often relatively inexpensive to secure, as are trade secrets. Patents, however, typically are not. It most often requires thousands of dollars to successfully acquire a patent. And because of a complex substantive examination process, patents also typically take years to receive. Depending on circumstances, however, patents are sometimes the best available mechanism for protecting many innovations.

To be eligible for patenting, an innovation must be patent eligible, new, and nonobvious in view of what has gone before. To be patent eligible means that the innovation must not be among the few categories of innovation that are not available for patenting regardless of how new and nonobvious. Excluded categories include such things as pure mathematical algorithms, natural phenomenon, and laws of nature, as well as financially abstract concepts. Also excluded are concepts that are completely lacking in utility, such as a perpetual motion machine.

The newness requirement specifies that the innovation differ in some way from that which has come before. In this example of a patented knife, where both knives and mirrors were previously known, it was new to combine the two (with the mirror at the end of the handle of the knife) (see Figure 21.2).

FIGURE 21.2
Example of patented knife.

In another example of an issued patent (see following figure), barcodes were previously known, as was the shape of Australia. It was new, however, to fashion a barcode in the shape of Australia.

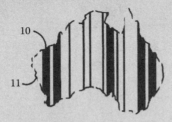

In addition to being different in some way or other from all previously known things or processes, the innovation must also be nonobvious. Obviousness is a thorny concept in patent law, and the word has little relationship to what the layperson might think it means. It will suffice for our present purposes to think of this requirement as follows: an innovation is obvious if a person of ordinary skill in the art would have some valid reason to combine two or more previously known concepts to thereby achieve the claimed innovation.

It is worth noting that patentability does not depend on how good an idea might be. That something has some modicum of utility, is new, and is nonobvious is not to necessarily also say that the idea is good. Conversely, many a good idea might not be categorically eligible for patenting, might not be new, or might be obvious. (Perversely, one might argue that the worse the idea, the less obvious it becomes, and hence the more patentable it is.) Accordingly, and contrary to how patents are often treated in the media and elsewhere, patents should not be thought of as miniature Nobel prizes.

All kinds of things and processes have been patented. One example is an issued patent for a painting kit (see Figure 21.3). Pursuant

FIGURE 21.3
Example of an issued patent.

to this innovation, one holds a baby, dips the baby's hindquarters in paint, and then daubs the baby's bottom on a canvas any number of times and using different paints to create a family heirloom of sorts.

Another example of an issued patent is titled "Method for creating anti-gravity illusion" (and happens to name the late entertainer Michael Jackson as an inventor) (see Figure 21.4). The idea here is to use a shoe having an appropriate notch formed therein that engages a nail protruding upwardly from the floor as an anchor to permit the performer to lean in an exaggerated manner.

And the patent in Figure 21.5 is titled "Method of exercising a cat." Here, the patent covers using a laser to create a spot that, when

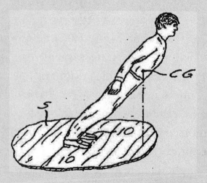

FIGURE 21.4
Example of an issued patent.

FIGURE 21.5
Example of an issued patent.

moved, entices a cat into a corresponding chase that entails exercise for the cat.

There are these and many, many more. The U.S. Patent and Trademark Office employs at the present nearly 9000 patent examiners who examine patent applications and assess whether to permit a corresponding patent to issue. Every week, the Patent Office issues thousands of new patents on an amazingly broad array of innovations.

GO for It

Pauline decides to invest in a patent application. After disclosing her information regarding her innovation to her patent attorney, the patent attorney works with her to draft a patent application. To some considerable extent, this patent application looks and reads like a technical white paper. There are illustrations and text that provide information about her process in significant detail.

TIME-OUT!

A patent application must disclose the invention in sufficient detail to permit a person of ordinary skill in this area of endeavor to make and practice the claimed invention without undue experimentation. In the United States, the inventor is also required to disclose the best mode known to the inventor for carrying out the invention. As a result, some inventors choose to avoid the patenting process in favor of maintaining their innovation as a trade secret.

There are also sentences called claims. An example of a claim from Pauline's patent application reads thusly:

1. A baking process comprising:
   - selecting a baking time
   - placing an edible material in an oven
   - initiating baking of the edible material
   - beginning at about halfway through the baking time, operating a Remora-brand vacuum cleaner within three to five feet of the edible material only for a period of time equal to about one-fourth the baking time

TIME-OUT!

A so-called utility patent application is required to have at least one claim, and may have as many more as the applicant may wish (keeping in mind that the patent office filing fee increases with each claim beyond 20 claims). A claim is a single sentence that describes that which the inventor has the right to prevent other people from making, using, or selling. Many patent claims read oddly and stilted to the layperson. This is sometimes partially owing to the inevitable technical jargon. More than this, however, is that patent claims are written in observation of a relatively unique syntax that has arisen over the years to (ironically enough) ensure clarity of meaning.

A series of patent claims often begin with a broad statement of the invention followed by a series of dependent claims that narrow that broad statement in some fashion. Note, for example, that Pauline's claim shown above is directed to an *edible material* rather than specifically a cookie. This is because her concept is potentially useful with other foodstuffs beyond cookies, and such a claim may help to include other items within the ambit of her patent. A dependent claim could then narrow that statement by specifying, for example, that the edible material comprises a cookie.

To be sure, a patent application represents considerable effort. The U.S. Supreme Court once observed, "The specification and claims of a patent, particularly if the invention be at all complicated, constitute one of the most difficult legal instruments to draw with accuracy." This effort, and the corresponding skill and expertise of the legal practitioner, unfortunately make pursuing a patent a relatively expensive proposition, especially compared to the other intellectual property rights.

Things Do NOT Move Quickly

It is nearly a year and a half before the Patent Office makes a substantive review and assessment of Pauline's patent application. In the meantime, the Titan Cookie Company essentially runs Pauline out of business.

TIME-OUT!

In fact, the U.S. Patent Office has a number of options by which an applicant can seek to speed up the examination process. Some of these options depend on the applicant's circumstances. For example, a petition to make an application special can be based on the applicant's senior status or the very poor state of their health. Another option permits the applicant to pay for fast-track treatment provided the applicant observes certain conditions and limitations with respect to the patent application and the number of claims being presented.

And that first substantive review, communicated in a so-called office action, amounts to a rejection of her patent application. The patent examiner assigned to her application has identified a number of prior art references that, according to the examiner, can be obviously combined to meet the features of each of her claims. Pauline and her patent attorney study this office action and the cited prior art references. Her attorney explains that they can leave the claims as they are and attempt to convince the examiner that the claims are in fact allowable over the examiner's position, or they can change one or more of the claims to better their position. Pauline is unimpressed with the examiner's position and decides to go for the argument-only option.

Her patent attorney prepares a formal response. In addition, the patent attorney conducts a telephone interview with the patent examiner to discuss the office action and Pauline's response. Pauline is fortunate, and the examiner decides to withdraw the rejections and to permit Pauline's patent application to be issued as a patent. While some patent applications are allowed relatively early during this so-called prosecution cycle, many patent applications must make their way through many rounds of rejections (requiring corresponding months or even years) before either issuing or the applicant abandoning the effort.

TIME-OUT!

Upon issuing, Pauline's patent will have a life measured from its original filing date. The usual term is 20 years from that filing date. This term is sometimes extended when the examination process is unduly delayed by the patent office. Pauline must also pay a maintenance fee once every four years (up to a maximum of three such payments) to keep her patent in force.

On the Offensive

While Pauline's patent application was making its way through the patent office, Titan's Cookie Company Xtreme cookie line became a $300 million-per-year product. Pauline brings suit against Titan for patent infringement. After some initial legal posturing, Titan offers to pay Pauline a royalty for further use of Pauline's patent. For Pauline, however, this whole situation has become very personal. She declines that offer and carries forward with the litigation.

TIME-OUT!

Were this an actual real-world example, we can be sure that Pauline's litigation counsel would be urging her to try to not take Titan's activities so personally, and to instead weigh the potential risks and benefits in as businesslike a manner as possible. No small part of that equation would be the often very high costs of pursuing a patent infringement lawsuit.

Four years later, we see that Pauline won at trial and later won at appeal. She has continued to ignore Titan's offer to pay a royalty and instead has succeeded in obtaining a court-ordered injunction that enjoins Titan from infringing her patent.

Recall, however, that Pauline's patent only protects a narrower version of her original concept. In particular, her patent only protects using the Remora-brand vacuum cleaner during the baking process. Titan, therefore, modifies its baking lines to use non-Remora vacuum cleaners and continues baking and selling cookies.

The public notices the difference in taste, however, and complains. Surprisingly quickly, Titan's Xtreme cookies are becoming the subject of jokes by late-night TV hosts.

Meanwhile, Pauline relaunches her website and begins selling the real deal. This time around, Pauline is marketing her cookies using a new name—*Got'cha Cookies*. Titan created a significant market demand for these cookies, and now Got'cha Cookies are the only way for consumers to satisfy that demand. Pauline's Got'cha Cookies begin selling the way you would expect a $300 million-a-year cookie brand to sell.

TIME-OUT!

Pauline's Got'cha expression is serving as a trademark. A trademark is basically any device, such as a word or logo, that indicates to a consumer a source or indication of quality. Trademarks are protected at common law but can also be registered at the state and/or federal level. A registered trademark can serve to help prevent others from using the same or confusingly similar device in their own marketing to thereby prevent consumers from being confused by unduly similar marks.

Getting It Right from the Start

Pauline files an application to register her new Got'cha trademark with the U.S. Patent and Trademark Office. Following an examination to ensure that her mark is not confusingly similar to another already-registered mark, and following publication of her mark to permit members of the public to object to her registration for some legitimate reason, her trademark is registered.

TIME-OUT!

A U.S.-registered trademark can potentially last indefinitely so long as the owner renews the registration every 20 years (and presuming that the owner, in fact, continues to make legitimate use of the mark in commerce).

Pauline Gets Older (and Wiser)

Pauline's patent eventually expires. The Titan Cookie Company is now free and clear to use Remora-brand vacuum cleaners when baking cookies. And so they do.

After all these years, however, Pauline's trademark Got'cha has garnered a great deal of consumer goodwill. It is a well-recognized brand with great consumer loyalty. This time, Pauline survives Titan's competition.

And, no, Pauline never did figure out exactly why the vacuum cleaner made a difference when baking her cookies.

FINAL TIME-OUT!

Pauline's story is not intended to exhaustively consider all possible best practices as regard intellectual property rights. Her story does not consider, for example, the opportunity to conduct a prior art search to assess the likelihood that she might be able to receive a patent before preparing and filing her patent application, or that one or more of her activities might encroach upon the intellectual property rights of someone else. Due-diligence practices often prompt business people to, for example, conduct trademark clearance searches or so-called freedom to operate studies to assess whether any third-party patents may present a legal obstacle to the manufacture and sale of a given product.

None of the various intellectual property rights is perfect. All are flawed in one way or the other. The various intellectual property rights are also not mutually exclusive. Their scope of protection can and does overlap with one another in many instances. A good approach to protecting intellectual property for a given business therefore often leverages more than one intellectual property right.

In Pauline's story, for example, her patent expires and her corresponding competitive advantage based on exclusivity disappears. By having built a worthy brand and protecting that brand via a trademark, a different kind of intellectual property right takes over from the expired patent and helps to protect the product line (albeit in a different way).

EXAMPLE

Examples are included in the portion of this chapter entitled "How to Use the Tool and Examples."

SOFTWARE

Essentially any decent word processing program will suffice to prepare suitable materials to express one's innovations in the context of seeking corresponding intellectual property rights.

WEBSITES

For patent searching:
- www.freepatentsonline.com—An excellent free online resource that provides access to both U.S. and a variety of non-U.S. patents and pending applications; to use this tool properly, it is worth spending some time becoming familiar with the *expert search* tool and capabilities.
- www.google.com/patents—Not an especially powerful search tool, but this free resource does give comprehensive treatment to all issued and pending U.S. patents.
- www.uspto.gov—In addition to providing searchable access to many issued and pending U.S. patents, there is a wealth of other information here regarding the operations of the U.S. Patent Office.

For federally registered trademark searching:
- www.uspto.gov

For copyright registration materials and information:
- www.copyright.gov

SUGGESTED ADDITIONAL READING

Gupta, P. and Trusko, B., eds. *Global Innovation Science Handbook*. Chapter 44: Intellectual property for innovation. New York: McGraw-Hill Professional, 2014.

22

Systems Thinking

Frank Voehl

CONTENTS

DEFINITION

Systems thinking is an approach to problem solving, by viewing *problems* as parts of an overall system, rather than reacting to specific part, outcomes, or events and potentially contributing to further development of unintended consequences.

USER

This tool can be used by individuals but its best use is with a group of four to eight people. Cross-functional teams usually yield the best results from this activity.

OFTEN USED IN THE FOLLOWING PHASES OF THE INNOVATIVE PROCESS

The following are the seven phases of the innovative cycle. An X after the phase name indicates that the tool/methodology is used during that specific phase.

- Creation phase X
- Value proposition phase X
- Resourcing phase X
- Documentation phase X
- Production phase X
- Sales/delivery phase X
- Performance analysis phase X

TOOL ACTIVITY BY PHASE

Systems thinking is a tool that can be used in all the phases of the innovation cycle when problems occur or when new approaches are being developed. It is particularly useful during the creation phase and the production phase.

HOW TO USE THE TOOL

Introduction to System Thinking

Systems thinking is the process of understanding how things influence one another within a whole. In nature, systems thinking examples include ecosystems in which various elements such as air, water, movement, plant, and animals work together to survive or perish. In organizations, systems consist of people, structures, and processes that work together to make an organization healthy or unhealthy.

Businesses are living organisms created to achieve goals. The paths leading to these goals are a complex web of local and global interdependencies resembling the neural structure of our brain. Organizations learn and grow by consuming and reacting to endless flows of information, recurring events, and forces involving group thought processes, emotions, and indelible cultural rhythms. Periods of poor and stellar performance wax and wane seemingly independent of rewarded accomplishments.

Nevertheless, they attempt control with new policies, improvement projects, and personal performance assessments in response to events and symptoms of corporate ill health, hoping to have an immediate measurable impact and sustained cure. They use fuzzy gut feelings, snapshot observations, and partial knowledge to form opinions about what causes the rise and fall of profit, quality, productivity, morale, and other interrelated performance variables. Everyone in the organization has a different and valid point of view but no means of organizing and utilizing this knowledge to help cure or reduce the impact of troubling corporate arrhythmia.

Systems thinking is not one thing but a set of habits or practices within a framework that is based on the belief that the component parts of a system can best be understood in the context of relationships with each other and with other systems, rather than in isolation. Systems thinking focuses on cyclical rather than linear cause and effect.

Systems thinking helps us move beyond technical solutions. While technology offers hope that they can build a more sustainable world, market failures limit the efficient allocation of capital and resources, including creativity and innovation. And there are long lags from problem recognition to innovation, commercial viability, and scale up. Technology often generates unintended consequences: for example, taller smokestacks reduce local smog but can often increase acid rain.

Innovation in markets, institutions, and governance is essential to realize the full potential of technology. Externalities must be priced. Market failures must be corrected. They can make technology more effective by improving market signals, through regulations that create level playing fields and prevent a race to the environmental bottom, and through monitoring to prevent free riding and unintended consequences.

Systems thinking helps us confront our values. Our guiding values offer the most important leverage point for enduring, sustainable change. Recently, I asked master of business administration students how much money they needed to be happy. The average response was $2 million per year, and about half said more is always better. Most would accept lower income—as long as they could make more than everyone else. But obviously, endless material growth on a finite world is impossible, and everyone cannot be richer than everyone else, no matter how clever our technology. Those who are currently affluent must confront the culture of consumption, the conflation of having with being, that is destroying both the environment and human well-being, while supporting the legitimate aspirations of billions around the world to rise out of poverty.

Systems thinking is very different from the way that most of us have learned to think. But if you embrace systems thinking, you may begin to see yourself surrounded by—and part of—many systems. Below are the simplified steps you should follow in order to begin systems thinking:

- *Step one*: Understand what systems are.
- *Step two*: Understand stocks, flows, and feedbacks.
- *Step three*: Apply these ideas to your company.

Step One: Many Systems Will Be Obvious, but the First Step to Thinking in a System Is Understanding Exactly How a System Works

Peter Senge and Donella Meadows, pioneers in systems thinking, define a system as a set of things—people, cells, molecules, or whatever—interconnected in such a way that they produce their own pattern of behavior over time. A key point here is that systems have inherent behaviors that produce certain results. All systems have three things:

- Elements
- Interconnections
- Functions *or* purpose

To illustrate this, Senge uses an example of a system of an old city neighborhood, in which people know one another and communicate regularly in a social system. The people, buildings, and stores are the *elements*; the relationships are the *interconnections*; and the *purpose* may be to create a safe environment to raise children, or just a sense of community. On the other hand, a brand new apartment block is not yet a system, not until new relationships form and a system emerges (i.e., not until the interconnections form).

Step Two: Understand Stocks, Flows, and Feedbacks

Once you see something as a system, the next step is to identify its *stocks*, *flows*, and *feedbacks*.

- *Stocks* are the foundations of systems. They are the elements that you can see, feel, count, or measure. It can be the water in a sink, the data on a server, the people at your company, or the money in your bank account.
- *Flows* are the actions that change the stocks of a system. For the examples above, the amount of water in a sink (stock) changes depending on how much water is poured in or drained out (flow); the money in your bank account (stock) changes as a result of your spending or saving (flow).
- *Feedbacks* or feedback loops, are mechanisms that create consistent behavior in a system. There are two kinds of feedback loops: *stabilizing loops* and *runaway (or reinforcing) loops*.
  - *Stabilizing loops*: If you are a coffee drinker, you can see your habit as a stabilizing loop. The stock in this case is your own energy level; the flow is the rate at which you get tired, or get your energy back. Since you cannot nap at work, as your energy declines, you drink more coffee to give your energy levels a boost. In other words, drinking coffee is meant to replenish your energy levels (stocks). Since the action is meant to keep your energy levels fairly even, this is a stabilizing loop.
  - *Runaway loops*: These loops can be good or bad. A simple example of this is your own bank account. The more money you have in there (stock), the more interest you will earn (flow). The more interest you earn, the more money gets deposited in your bank account. And so on, and so on. This loop will continue until a different behavior is introduced.

Step Three: Apply These Ideas to Your Company

Begin to think about how your own company is a system, and how existing loops may be influencing behavior. For example, how is the company culture influencing behavior? And how is that behavior then influencing the culture?

Advantages

1. Systems thinking can help us see that *failing* systems may really simply be designed for a purpose other than what we assume or have been told.
2. It is specifically designed to look at the dynamics of a system quantitatively, even though it may be based on a qualitative analysis.
3. You can test and develop hypotheses.
4. The debates around the relationships between the systems components usually exposes deeply held but often untested assumptions about the way the system works.
5. Systems thinking can help us see that what may seem an isolated problem is actually part of an interconnected network of related issues.
6. Systems thinking can help us see the positive and negative feedback cycles that may be affecting an issue of importance to us. Feedback cycles may involve delays. It is important to be aware in these and other systems, as we attempt to optimize them, and Systems thinking teaches us that feedback does not always happen instantly.
7. Systems thinking can help us remain aware of the time delays between the onset and effects of feedback relationships. Attempting to solve complex issues without a systems thinking approach may lead to unintended consequences, despite our best intentions.

Disadvantages

1. One of the biggest drawbacks in trying to implement systems thinking techniques is making the assumption that everyone else naturally sees the problems the way we see them, and therefore can easily switch to using our viewpoint (tools and concepts) to address the problem.

2. If you do not go the simulation route (i.e., you map the system but do not model it), you cannot easily explore how all the variables interact with each other dynamically.

3. Not only are we unsure that we have identified the most important variables, but we also do not know for certain the actual scale of their relative dynamics within the system.

4. Some people get obsessed with the need to include absolutely every possible variable, essentially confusing simulation of reality with modeling to explore and expose assumptions.

5. In deeply qualitative situations, the task of getting accurate and valid data for modeling can require a great deal of expertise and can be very time consuming.

6. Often, entirely new information systems need to be set up to collect the data.

7. According to some experts like Bob Flood, it does not model emergent phenomena very well, although it may well model the emergent process.

8. It does not grapple very well with the problem of what to do when the whole thing expands beyond the realm of investigation.

9. There are no hard and fast rules for boundary setting.

10. It is not inherently *reflective*, although most people would use some form of reflective process in assessing what the model says.

SUMMARY

Systems thinking is a powerful set of problem-solving tools and techniques based on system analysis and design (explained by Peter Senge in *The Fifth Discipline*) that helps us avoid unintended consequences and find optimal solutions to complex problems. Systems thinking is a philosophy that looks at the world in terms of just what it says—systems. The entire world can be seen as one big system that encompasses countless smaller systems.

Systems thinking identifies the elements of a system as feedback cycles and delays. By explaining how these cycles and delays work to create and change the systems around us—from our economic and political systems right down to the human systems made up of and inside our own bodies and minds—systems thinking offers an incredible set of problem-solving tools and techniques to help us understand and optimize areas suffering due to complex problems. Also, it helps us to recognize constraints.

Many of us are overstressed and operate in overstressed organizations. Trying to do too much means they are often unable to marshal the resources they need to kick-start improvements in productivity, quality, and sustainability. The result is a self-reinforcing trap of low-performance, overstressed resources and failed improvement programs. Firms that succeed in quality and sustainability free up the resources needed to improve by slowing down and focusing on the long term.

Finally, it helps us to recognize that they can make a difference. People often feel powerless in the face of huge, complex systems. But understanding how systems work helps us find the high leverage points that make a difference.

EXAMPLES

1. More and more businesses are developing the systems thinking capabilities of their people, and realizing large benefits. For example, using systems thinking

   - A major oil company has generated documented savings of several billion dollars to date, while improving safety and environmental quality.
   - A shipyard went from cost overruns and project delays to an award-winning yard in great demand.
   - Businesses bootstrap steady improvement in quality, productivity, and sustainability by reinvesting initial savings in further improvement.
   - A high-technology electronics firm redesigned its supply chain, improving customer service and delivery reliability while cutting inventory.
   - A global automaker built an entirely new service business and is now the market leader in that rapidly growing segment.
   - A major university implemented maintenance projects that boosted energy efficiency and sustainability while more than paying for themselves, creating resources for still more projects.

2. Another good example of how systems thinking can apply to a company is what is called *drift to low performance.* Many companies measure performance against past results. If an employee boosts his or her sales compared with last year, he or she is rewarded. If sales have been declining, though, even slight improvements are seen as positive. However, this degradation in performance then leads to a degradation in company goals, as the yardstick gets longer and longer. There are two antidotes to this type of *runaway loop.* Keep standards absolute regardless of performance, and make goals sensitive to only the best performance of the past, not just the past year or poor performance.

3. How systems thinking can benefit a business can be found in energy efficiency. Upgrading light bulbs one at a time may not seem worth it, but savings can really add up when you tackle multiple projects at once, as improvements in one area can positively affect other areas. For example, sealing leaks in your air ducts can not only make your heating, ventilation, and air conditioning system more efficient but also lower your electric bills while making your office more comfortable for your employees. More comfortable employees may be more productive, which can increase their performance, which can boost revenues, and more revenues can fund retrofits for a more comfortable office.

CASE STUDY

The Lean Enterprise Research Center in Wales applied systems thinking to the public sector in 2010, and carefully tracked performance. Systems thinking can be powerful, but too often remains an abstract concept instead of a useful tool. The challenge is to develop our systems thinking skills, to help others develop their capabilities, and to bring systems thinking into everyday life: to move beyond slogans and into real action.

A disabled care service faced very high demand, and very slow end-to-end times in delivering its service. With an external team of consultants, the facility took a long look at all the elements in its system—its employees and training, its physical facilities, its processes, etc.—as well as the interactions between elements. It took some time for all employees to embrace this new way of thinking about how to perform daily operations, but as people came on board with systems thinking the organization was able to make deep changes to its services. As a result of the changes made to the system, which were also changes to the underlying behaviors of the system, the organization saw improvements across the board.

A key point in this case is that employees began by trying to speed up the rate at which they processed customers. But after applying systems thinking, they began to address how they could prevent some of those needs in the first place, which resulted in much faster end-to-end times. In other words, before systems thinking, they were trying to treat the symptoms; after, they treated the causes.

SOFTWARE

- *isee systems* (formerly High Performance Systems): Their business model is to improve the way the world works, by creating systems thinking–based products that enable people to increase their capacity to think, learn, communicate, and act more systemically. These software products include STELLA, °iThink, °NetSim, Modeler, Player, XMILE, and others. For further information, see http://www.iseesystems.com/AboutUs.aspx.
- *Vensism*: Their business model is to provide industrial-strength simulation software for improving the performance of real systems. Vensim's rich feature set emphasizes model quality, connections to data, flexible distribution, and advanced algorithms. Configurations for everyone from students to professionals. See http://vensim.com/.
- *Insight Maker*: Insight Maker runs in your web browser. No software download or plug-ins are needed. Get started building your rich pictures, simulation models, and Insights. For further information, see http://insightmaker.com/.
- *AnyLogic*: A multimethod simulation software. The only simulation tool that supports discrete event, agent based, and system dynamics simulation. AnyLogic supports the design and simulation of feedback structures (stock and flow diagrams and decision rules, including array variables also known as *subscripts*) in a way most system dynamics modelers are used to. See http://www.anylogic.com/system-dynamics.

REFERENCE

Senge, P.M. *The Fifth Discipline: The Art & Practice of the Learning Organization.*

SUGGESTED ADDITIONAL READING

Ackoff, R.L. *Re-creating the Corporation: A Design of Organizations for the 21st Century.*

Ackoff, R.L. and Addison, H.J. *Systems Thinking for Curious Managers: With 40 New Management f-Laws.*

Boardman, J. and Sauser, B. *Systems Thinking: Coping with 21st Century Problems.*

Checkland, P. *Systems Thinking, Systems Practice: Includes a 30-Year Retrospective.*

Gharajedaghi, J. *Systems Thinking, Second Edition: Managing Chaos and Complexity: A Platform for Designing Business Architecture.*

Haines, S.G. *The Manager's Pocket Guide to Systems Thinking and Learning.*

Jackson, M.C. *Systems Thinking: Creative Holism for Managers.*

Laszlo, E. *The Systems View of the World: A Holistic Vision for Our Time.*

Meadows, D. *Thinking in Systems: A Primer.*

Richmond, J., Stuntz, L., Richmond, K., and Egner, J. *Tracing Connections: Voices of Systems Thinkers.*

Sherwood, D. *Seeing the Forest for the Trees: A Manager's Guide to Applying Systems Thinking.*

Sterman, J.D. *Business Dynamics: Systems Thinking and Modeling for a Complex World.*

Sweeney, L.B. and Meadows, D. *The Systems Thinking Playbook: Exercises to Stretch and Build Learning and Systems Thinking Capabilities.*

Weinberg, G.M. *An Introduction to General Systems Thinking.*

23

Value Proposition

H. James Harrington

CONTENTS

The best time to stop a marginal project is before you start it. Most organizations have lots of ideas available to them but few of them are worth investing in.

H. James Harrington

DEFINITION

A *value proposition* is a document that defines the benefits that will result from the implementation of a change or the use of an output as viewed by one or more of the organization's stakeholders. A value proposition can apply to an entire organization, parts thereof, or customers, or products, or services, or internal processes.

A *business case* captures the reasoning for initiating a project or task. It is most often presented in a well-structured written document, but in some cases may come in the form of a short verbal agreement or presentation. A business case is preferably prepared by an independent group after the project concept has been approved by executive management.

USER

This tool can be used by individuals but its best use is with a group of four to eight people. Usually, the best results from preparing a value proposition is realized when the group is made up of members from cross-functional units.

OFTEN USED IN THE FOLLOWING PHASES OF THE INNOVATIVE PROCESS

The following are the seven phases of the innovative cycle. An X after the phase name indicates that the tool/methodology is used during that specific phase.

- Creation phase
- Value proposition phase X
- Resourcing phase X
- Documentation phase
- Production phase
- Sales/delivery phase
- Performance analysis phase

TOOL ACTIVITY BY PHASE

- Value proposition phase—During the value proposition phase, the value proposition is prepared and presented to the executive committee or the organization assigned the responsibility for approving new projects.
- Resourcing phase—During the resourcing phase, the value proposition is used to prepare the business case that is used to justify financing the project. Also during this phase, other resources like staffing, facilities, and equipment is acquired.

HOW TO USE THE TOOL

Affinity value propositions are prepared to determine if the proposed project will have a positive or negative value added to the major stakeholders. It takes into consideration all of the costs related to managing and implementing the project, such as the cost of operating the affected systems/processes compared with the value added in continuing to operate without making any changes. In most cases, the organization needs to consider how accurate these estimations are in making a decision to approve or disapprove the proposed project.

A cost–benefit analysis (CBA) is a critical part of a value proposition. A CBA is anything used to compare the benefit of a proposed process change with the cost associated with implementing that change. The higher the ratio of the measurable benefits to the quantifiable costs, the easier it is to spend resources on a proposed process implementation. In the simplest form of CBA, there are two primary measures:

- Positive financial benefit = financial impact – the costs of implementation
- Payback = implementation costs/annual benefits

When doing a CBA, there are some key principles that need to be considered:

- Use the same units of measurement for both costs and benefits.
- Compare the situation with and without the process change to assess its real impact.
- Consider the specific location of the study (and its unique characteristics) for transferring the results to the other area.
- Avoid double counting either benefits for costs.

Steps in preparing a value proposition:

1. Define the costs related to the value proposition group's activities.
2. Estimate the cost related to implementing the proposed project.
3. Estimate the time required to implement the proposed project.
4. Define the risk related to implementing the proposed project. Define both the positive and negative impacts the project will have on major stakeholders (both tangible and intangible).

5. Define assumptions that the estimates are based on.
6. Prepare the value proposition report.
7. Present the value proposition report to executive management.

EXAMPLE

The following is a typical example of the table of contents for a typical value proposition report.

- Executive overview
- Description of current state
- Added value that the proposed change would produce
- Description of the propose change
- Backup data
- Costs and required time frame
- Other solutions considered
- Risk and obstacles
- Recommendations
- Key individuals
- Financial calculations
- Other value-added results—both tangible and intangible
- Lists of assumptions
- Implementation plan

Note: Be sure that you make your point in the executive overview. The rest of the report only serves as backup. Some executives will not have the time to read the full report and will make their decision based on the executive overview.

SOFTWARE

Some commercial software available includes but is not limited to

- WinSite: http://value-proposition.winsite.com
- Logix Guru: http://www.logixguru.com/

SUGGESTED ADDITIONAL READING

Palomino, J. *Value Prop—Create Powerful I3 Value Propositions to Enter and Win New Markets*. Philadelphia: Cody Rock Press, 2008.

Hardy, J.G. *The Core Value Proposition: Capture the Power of Your Business Building Ideas*. Cheshire, UK: Trafford Publishing, 2005.

Harrington, H.J. and Trusko, B. *Maximizing Your Value Proposition*. Boca Raton, FL: Taylor & Francis, 2014.

Gyorffy, L. and Friedman, L. *Creating Value with CO-STAR: An Innovation Tool for Perfecting and Pitching Your Brilliant Idea*. Palo Alto, CA: Enterprise Development Group, 2012.

24

Vision/Goals

H. James Harrington

CONTENTS

DEFINITION

- Vision—A documented or mental description or picture of a desired future state of an organization, product, process, team, a key business driver, activity, or individual.
- Vision statements—A group of words that paints a clear picture of the desired business environment at a specific time in the future. Short-term vision statements usually are from three to five years. Long-term vision statements usually are from ten to 25 years. A visions statement should not exceed four sentences.
- Goal—The end toward which effort is directed: the terminal point in a race. These are always specified in time and magnitude so they are

easy to measure. Goals have key ingredients. First, they specifically state the target for the future state and, second, they give the time interval in which the future state will be accomplished. These are key input to every strategic plan.

USER

One of the primary responsibilities of the innovator is to ensure that aggressive but realistic organizational/product vision and goals are established and communicated throughout the organization. In finalizing vision and goals, the innovator should involve as many of the top management team as possible so that he or she will have their support.

OFTEN USED IN THE FOLLOWING PHASES OF THE INNOVATIVE PROCESS

The following are the seven phases of the innovative cycle. An X after the phase name indicates that the tool/methodology is used during that specific phase.

- Creation phase X
- Value proposition phase X
- Resourcing phase X
- Documentation phase X
- Production phase X
- Sales/delivery phase X
- Performance analysis phase X

TOOL ACTIVITY BY PHASE

- Creation phase—During the creation phase, the innovator or the creative team will develop value statements that define how the innovative product, service, or situation will affect the customer/receiver of the innovative activity. Goals are established to quantify expected impact, timing, and value added to the innovative organization as a result of implementing the innovative concept.

- Value proposition phase—During the analysis and preparation activities related to preparing value proposition, the vision of who will be affected and how they will be affected are refined and updated. The goals are also evaluated and updated to reflect the information that has been collected. These visions and goals are one of the primary inputs to the value proposition.

- Resourcing phase—The comparison of the financial plan and the vision and goals for the innovation concept must be supportive of each other. Often the optimism that is reflected in the vision and goals is completely out of line with the actual financial requirements to implement the innovation concept. This is a critical phase where many of the innovative concepts are abandoned because adequate funds cannot be obtained internally or externally. Investors can become very enthusiastic about a vision and a supporting set of goals but quickly become disillusioned when the actual financial impacts are defined.

- Documentation phase—As the organization documents the innovation implementation plan and requirements, it must consider the vision and the goals established for the innovation concept. The formal documentation and the vision and goals must be in harmony in order to eliminate confusion and to have any hopes of having a successful program.

- Production phase—The goals set related to output performance, costs, and delivery schedules are a primary input in defining how the production activities are organized, and the quality of the production process, and the timing of delivery to external customers.

- Sales/delivery phase—The sales and marketing strategy/activities must be based on the vision statement that defines the potential receivers of the outputs from the innovative process. Sales quotas, market segments, pricing, and quantity targets are derived from the goals set related to the outputs created by the innovative process.

- Performance analysis phase—The performance analysis activities compare actual results to the established vision and goals for the innovative concept. Success or failure is measured by the degree the visions and goals established for the innovative activity met or exceeded when the actual results are analyzed. Having realistic and attainable visions and goals is absolutely essential for a successful innovative activity.

HOW TO USE THE TOOL

Jack Welch, former CEO of General Electric, stated, "Leaders—and you can take anyone from Roosevelt to Churchill to Reagan—inspire people with clear visions of how things can be done better. Sometimes managers, on the other hand, muddle up things with pointless complexity and detail. They equate (managing) with sophistication, with sounding smarter than anyone else. They inspire no one."

> The very essence of leadership is that you have to have a vision. It's got to be a vision that you articulated clearly and forcefully on every occasion. You cannot blow an uncertain trumpet.
>
> **Father Theodore Hesburgh**
> *former president of the University of Notre Dame*

All too often, organizations prepare unrealistic vision statements using phrases like *the biggest, the largest,* and *the most profitable.* Each of these phrases can apply to only one organization in the organization's field of competency. Much better are phrases such as *world class, recognized as a leader,* and *one of the best.*

Organizations that excel set high standards for themselves and their employees. The following are some best practice guidelines for developing vision statements.

- A vision statement is a major, valuable, ambitious goal that clearly describes a desirable future achievement at a particular moment.
- Management inspires commitment to a clear vision and defined goals.
- The vision is communicated to everyone in such a way that all know it and understand it.
- The vision is sufficiently ambitious and meaningful to enable everyone to contribute to its implementation.
- The vision penetrates the life of the organization, mobilizes energy, and creates a passion for success.

President John F. Kennedy gave a good example of a convincing and challenging vision when he declared that the United States would send a man to the moon and bring him back safely at the end of the 1960s.

A vision statement should be prepared by the individuals who are most affected by the future state conditions. Organizational and product vision statements are usually prepared by top management and directed at what the organization's or specific product's output will be like, or how it will be used at a specific point in time in the future. Vision statements can be long-term or short-term statements. They are key parts of most business plans.

SHORT-TERM VISION

A short-term vision is the image of an entity's future—the direction it is headed, the customer focus it should have, the market position it should try to occupy, the organization's activity in the pursuit, and the capability plan that has been developed. A short-term strategic vision should delineate what kind of organization is trying to become at the end of a strategic time frame, which is typically three to ten years in the future. This will infuse the organization with a sense of purposeful action.

A short-term strategic vision statement for an organization should answer the following questions:

- What are the key factors underlining the current situation that will influence the way the organization should look like five years from now?
- What assumptions will drive the needed transformation?

In preparing a short-term vision, we find it is often useful to start with a SWOT (strength, weaknesses, opportunities, and threats) analysis that will help you look at your organization's mission, vision, and goals. This approach will help you answer the following questions:

- What do we do best?
- What do we not do best?
- What are our organization's resources—asset, intellectual properties, and people?
- What is our organization's capability (function)?
- What is happening externally that will affect our organization?
- What are the strengths and weaknesses of each of our competitors?

- What are the driving forces behind our sales trends?
- What are the important and potentially important markets?
- What is happening in the world that might affect us?

The major features of an effective short-range vision statement include

- Clarity and lack of any ambiguity
- Valid and clear picture
- Description of a brighter future
- Memorable and engaging wording
- Realistic aspirations
- Alignment with organization's value and culture

Visions and Goals That Innovators Use

We will focus this part of the chapter on how the professional innovator should be using visions and goals to accomplish his or her objectives. The professional innovator can be involved in four different types of visions/goals:

1. Personal
2. Organizational
3. Product/innovative concept
4. Activity

Personal Visions and Goals

Every professional innovator should have a set of both short- and long-range personal vision statements and goals that are to be used to set direction and measure progress. A typical personal vision statement might read like the following: "Develop a sufficient number of innovative ideas and guide them through the innovative process generating sufficient personal wealth to allow me to retire and live comfortably at the age of 50." A typical personal goal that would support this vision statement could read: "Accumulate personal hard assets that could be liquidated at any time that exceeds $4.5 million (after all the debts I have outstanding are paid off) as a result of the innovative concepts that I personally created and implemented." As you can see, personal vision and goals are developed to set the direction and focus of an individual's personal career and personal activities.

Organizational Visions and Goals

Organizational vision and goals are developed to set the direction and activities of the organization that a professional innovator establishes to implement one or more of his or her innovative concepts. In this case, the innovator and his or her executive team is responsible for creating, updating, and communicating these key directives. In other cases, the organizational vision and goals are set by the executive team of the organization that the innovator is part of. For example, these could be the visions and goals for a company like IBM that sets the expectations, type of creative solutions developed, and defines the role that the innovator plays within the organization. For example, this could define the role that the personnel in research and development (R&D) play within IBM or any other organization that hires people that make up parts of the innovation process. In this case, a typical vision statement could read: "To be recognized worldwide as one of the leaders in developing new and creative technology related to the telecommunications industry." A typical goals statement that would support this vision statement could read: "A minimum of 52% of our annual income would be generated by new products/services that were not part of our offerings three years prior."

PRODUCT/INNOVATIVE CONCEPT VISION AND GOALS

These are vision and goal statements that support an individual and unique activity. They relate to an individual creative unique idea that has the potential of completing the total innovative process. They can be an individual and unique idea that is the basis for forming a new organization, or they can be one of a series of innovative concepts that are designed to penetrate a market structure that has already been established. In this case, they are usually new programs that are approved as a present product starts down the negative slope of the S-curve. In this case, a typical vision statement could read: "The new 12N module will provide more value-added to its users at a lower price and higher reliability thereby obsoleting the present 03M module." The goal statement to support this vision statement might read: "The new 12N module should have 50% more capacity, be 30% smaller, be equally as reliable, and sell for 20% less than the present 03M module."

Activity Vision and Goals

In most organizations, the innovative process is divided into major functional units like R&D, finance, production, product engineering, project management, manufacturing engineering, quality assurance, and sales and marketing. In these cases, the individuals that make up these functional units play a key role in the innovative process. As a result, their visions and goals must reflect the role that they are required to excel at in order for the innovative process to work efficiently and effectively. For example, a typical innovative vision statement for the personnel in sales and marketing might read like: "We will employ individuals that develop unique and creative marketing concepts that distinguish our organization's products and services from those provided by the competition." A typical goal that would support this vision statement might read like: "Our marketing strategy/activities cost will be no greater than 5% of the revenue generated as a result of leads related to the marketing campaigns. Our website hit rate will be a minimum of 1000 individuals per day that request follow-up contact."

EXAMPLE

A number of examples of vision and goals statements have already been presented in this chapter. The following are some specific examples of mission and goals statements by major organizations.

- Ford's vision statement: "To become the world leading consumer company for automotive products and/or services."
- CNN's long-term vision statement: "To provide first-hand news from all around the world as it happens, customized to each countries interests and needs."
- Cleveland Clinic: "Striving to be the world's leader in patient experience, clinical outcomes, research and education."
- Creative Commons: "Our vision is nothing less than realizing the full potential of the Internet—universal access to research and education, full participation in culture—to drive a new era of development, growth, and productivity."

The following was Toyota's global short-term (eight years) vision statement: "innovation into the future—a passion to create a better society through manufacturing of value-added products and technological innovation." Toyota's backup strategy to support this vision statement was

- Be a driving force in global regeneration by implementing the most advanced environmental technologies
- Create automobiles and motorized society in which people can live safely, securely, and comfortably
- Promote the appeal of cars throughout the world and realize a large increase in the number of the Toyota fans
- Be a global company that is trusted and respected by all people around the world

Examples of goal:

When IBM was faced with a significant group of competitors that were addressing their marketplace, its management team developed a ten-year goal strategy. It stated that in the next ten years IBM would

- Grow with the industry
- Exhibit product leadership across the entire product line
- Excel in technology, value, and quality
- Be the most effective in everything we do
- Be the low-cost provider, low-cost seller, and low-cost administrator
- Sustain our profitability; funds are growth

Motorola, when faced with significant competitive pressure, established a five-year goal program. Motorola's CEO William J. Weisz issued the following directive to the total organization: "We developed as one of our top ten goals for the company, the Five-Year Tenfold Improvement Program. This means that no matter what operation you are in, no matter what your present level of quality performance, whether you are a service organization or manufacturing organization, it is our goal to have you improve that level by an order of magnitude in five years."

One of the very few really aggressive improvement goals that our national government has placed on its operations was issued by President Ronald Reagan when he released Executive Order 12552, stating, "There is hereby established a government-wide program to improve the quality, timeliness, and effectiveness of service provided by the federal government. The goal of the program shall be to improve the quality and timeliness of service to the public, and to achieve a 20% productivity increase in appropriate functions by 1992. Each executive department and agency will be responsible for contributing to the achievement of this goal."

It is just too bad that this goal was not enforced by the follow-on management system within our federal government.

SOFTWARE

- GoalEnforcer: http://www.goalenforcer.com
- Goalscape: http://www.goalscape.com
- Goal Buddy: http://www.goal-buddy.com
- Goal Genius 2012—Inspired Pursuits: http://goalgenius.zendesk.com
- Writing S.M.A.R.T. Goals—Digital Software Development
- Goals on Track: http://www.goalsontrack.com

SUGGESTED ADDITIONAL READING

Angelica, E. *Crafting Effective Mission and Value Statements*. St. Paul, MN: Wilder Publishing Center, 2002.

Evans, D.A. *The Vision Statement*. Raleigh, NC: LULU-Rodbeaer, 2012.

Harrington, H.J. and Harrington, J.S. *Total Improvement Management—The Next Generation in Performance Improvement*. New York: McGraw-Hill, 1995.

Harrington, H.J. and Voehl, F. *The Organizational Alignment Handbook—A Catalyst for Performance Acceleration*. Boca Raton, FL: CRC Press, 2012.

Harrington, H.J. and Voehl, F. *The Organizational Master Plan Handbook—A Catalyst for Performance Planning and Results*. Boca Raton, FL: Taylor & Francis LLC, 2012.

Appendix: Innovation Definitions

The following terms and definitions are all direct quotes from the International Association of Innovation Professionals (IAOIP) "Study Guide to the Basic Certification Exam." The study guide is used to prepare individuals to take the basic examination to be certified as a professional innovator by the IAOIP. These were taken from what the IAOIP is using as the body of knowledge for the innovation professional.

Notes:

1. The terms and definitions that are italicized were not included in the "Study Guide to the Basic Certification Exam" but are included in the methodologies and tools documented technology.
2. In some cases, there is more than one definition for the same tool/methodology. In most of these cases, all the definitions are acceptable and often the additional definitions just help to clarify what the methodology/tool is. In some cases, the preferred definition is identified.

TERMS AND DEFINITIONS

5 Whys: *A simple but effective method of asking five times why a problem occurred. After each answer, ask why again using the previous response. It is surprising how this may lead to a root cause of the problem, but it does not solve the problem.*

5 Whys: A technique to get to the root cause of the problem. It is the practice of asking five times or more why the failure has occurred in order to get to the root cause. Each time an answer is given, you ask why that particular condition occurred. As outlined in the 5 Whys Overview, it is recommended that the 5 Whys be used with Risk Assessment in order to strengthen the use of the tool for innovation and creativity-enhancing purposes.

76 Standard solutions: *A collection of problem-solving concepts intended to help innovators develop solutions. A list was developed from referenced works and published in a comparison with the 40 principles to show that those who are familiar with the 40 principles will be able to expand their problem-solving capability. They are grouped into five categories as follows:*

1. *Improving the system with no or little change 13 standard solutions*
2. *Improving the system by changing the system 23 standard solutions*
3. *System transitions six standard solutions*
4. *Detection and measurement 17 standard solutions*
5. *Strategies for simplification and improvement 17 standard solutions*

7–14–28 Processes: This is a task-analysis assessment that involves breaking a process down into seven tasks, then breaking it further into 14 tasks, and then another level further into 28 tasks.

40 Inventive principles: The 40 inventive principles that form a core part of the TRIZ (theory of inventive problem solving) methodology invented by G.S. Altshuller. These are 40 tools used to overcome technical contradictions. Each is a generic suggestion for performing an action to, and within, a technical system. For example, principle #1 (segmentation) suggests finding a way to separate two elements of a technical system into many small interconnected elements.

AEIOU frameworks: This is a way to observations, and stands for activities, environments, interactions, objects, and users. It serves as a series of prompts to remind the observer to record the multiple dimensions of a situation with textured focus on the user and their interactions with their environment.

ARIZ (algorithm for creative problem solving): A procedure to guide the TRIZ student from the statement of the IFR (ideal final result) to a redefinition of the problem to be solved and then to the solutions to the problem.

Absence thinking: *Absence thinking involves training the mind to think creatively about what it is thinking and not thinking. When you are thinking about a specific something, you often notice what is not there, you watch what people are not doing, and you make lists of things that you normally forget.*

Abstract rules: Abstract rules are those unarticulated, yet essential, guidelines, norms, and traditions that people within a social setting tend to follow.

Abundance and redundancy: Abundance and redundancy is based on belief (not necessarily factual) that if you want a good invention that solves a problem, you need lots of ideas.

Administrative process: This specifies what tasks need to be done and the order in which they should be accomplished, but does not give any, or at least very little, insight as to how those tasks should be realized.

Affinity diagram: Affinity diagram is a technique for organizing a variety of subjective data into categories based on the intuitive relationships among individual pieces of information. It is often used to find commonalties among concerns and ideas. It lets new patterns and relationships between ideas be discovered.

Affordable loss principle: It stipulates that entrepreneurs risk no more than they are willing to lose.

Agile innovation: This is a procedure used to create a streamlined innovation process that involves everyone. If an innovation process already exists, then the procedure can be used to improve the process resulting in a reduction of development time, resources required, costs, delays, and faults.

Analogical thinking and mental simulations: Using past successes applied to similar problems by mental simulations and testing.

Application of technology: These people are intrigued by the inner workings of things. They may be engineers, but even if not, they like to analyze processes, get under the hood, and like to use technology to solve problems (business or technical).

Architect: Designs (or authorizes others to design) an end-to-end, integrated innovation process, and also promotes organization design for innovation, where each function contributes to innovation capability.

Attribute listing, morphological analysis, and matrix analysis: Attribute listing, morphological analysis, and matrix analysis techniques are good for finding new combinations of products or services. We use attribute listing and morphological analysis to generate new products and services. Matrix analysis focuses on businesses. It is used to generate new approaches, using attributes such as market sectors, customer needs, products, promotional methods, etc.

Attributes-based questions: Questions based on attributes are ones in which you look for a specific attribute of an object or idea.

Balanced breakthrough model: This suggests that successful new products and services are desirable for users, viable from a business perspective, and technologically feasible.

Barrier buster: This helps navigate political landmines and removes organizational obstacles.

Benchmark (BMK): *The standard by which an item can be measured or judged.*

Benchmarking (BMKG): *A systematic way to identify, understand, and creatively evolve superior products, services, designs, equipment, processes, and practices to improve your organization's real performance.*

Benchmarking innovation: A form of contradiction. Doing something completely new—applying an invention in a new way. It means that others are not doing the same thing. Thus, there is nothing to benchmark.

Biomimicry: *Biomimetic or biomimicry is the imitation of the models, systems, and elements of nature for the purpose of solving complex human problems (Wikipedia). It is the transfer of ideas from biology to technology; the design and production of materials, structures, and systems that are modeled on biological entities and processes. The process involves understanding a problem and observational capability together with the capacity to synthesize different observations into a vision for solving a problem.*

Bottom–up planning for innovation: A process where innovations are described in portfolio requirements to meet business objectives.

Brainstorming: *A technique used by a group to quickly generate large lists of ideas, problems, or issues. The emphasis is on quantity of ideas, not quality.*

Brainstorming or operational creativity: Brainstorming combines a relaxed, informal approach to problem solving with lateral thinking. In most cases, brainstorming provides a free and open environment that encourages everyone to participate. While brainstorming can be effective, it is important to approach it with an open mind and a spirit of nonjudgment.

Brainwriting 6–3–5: *An organized brainstorming with writing technique to come up with ideas in the aid of innovation process stimulating creativity.*

Breakthrough, disruptive, new-to-the-world innovation: Paradigm shifts that reframe existing categories. Disruptive innovation drives significant, sustainable growth by creating new consumption occasions and transforming or obsolescing markets.

Bureaucratic process: This occurs where the inputs are defined and a specific routine is performed; however, the desired output is obtained only by random chance.

Business case: *A business case captures the reasoning for initiating a project or task. It is most often presented in a well-structure written document, but in some cases may come in the form of a short verbal agreement or presentation.*

Business Innovation Maturity Model (BIMM): This offers a road map to innovation management maturity.

Business model generation canvas: This is a tool that maps what exists. The business model canvas is a strategic management and entrepreneurial tool comprising the building blocks of a business model. The business is expressed visually on a canvas with the articulation of the nine interlocking building blocks in four cluster areas: offering—value proposition; customer—customer segments, customer relationships; infrastructure—distribution channels, key resources, key partnerships, key activities; value—cost structure and revenue model.

Business model innovation: This changes the method by which an organization creates and delivers value to its customers and how, in turn, it will generate revenue (capture value).

Business plan: *A business plan is a formal statement of a set of business goals, the reason they are believed to be obtainable, and the plan for reaching these goals. It also contains background information about the organization or team attempting to reach these goals.*

CO-STAR: *This is specifically designed to focus the creativity of innovators on ideas that matter to customers and have relevance in the market. It is an easy-to-use tool for turning raw ideas into powerful value propositions.*

Capital investment: This is the cost of manufacturing equipment, packaging equipment, and change parts.

Cause-and-effect diagram: *A visual representation of possible causes of a specific problem or condition. The effect is listed on the right-hand side, and the causes take the shape of fish bones. This is the reason it is sometimes called a "fishbone diagram" or an "Ishikawa diagram."*

Co-creation innovation: A way to introduce external catalysts, unfamiliar partners, and disruptive thinking into an organization in order to ignite innovation. The term co-creation innovation can be used in two ways: co-development and the delivery of products and services by two or more enterprises; and co-creation of products and services with customers.

Collective effectiveness: In a complex and highly competitive business environment, it is difficult to sustain or support research and development (R&D) and innovation expenses. Networking allows firms access to different external resources like expertise, equipment, and overall know-how that has already been proven with less cost and in a shorter period.

Collective learning: Networking not only helps firms gain access to expensive resources like machinery, laboratory equipment, and technology, but it also facilitates shared learning via experience and good practice sharing events. This brings new insight and ideas for a firms' current and future innovation projects.

Combination methods: A by-product of already applied process, system, product, service wise solutions integrated into a one solution system to produce one end-result that is unique.

Communication of innovation information: Employees vary greatly in their ability to evaluate potentially significant market information and convey qualified information to pertinent receivers in the product development stream.

Comparative analysis: A detailed study/comparison of an organization's product or service to the competitors' comparable product and or service.

Competitive analysis: It consists of a detailed study of an organization's competitor products, services, and methods to identify their strengths and weaknesses.

Competitive shopping (sometimes called mystery shopper): This is the use of an individual or a group of individuals that goes to a competitor's facilities or directly interacts with the competitor's facilities to collect information related to how the competitor's processes, services, or products are interfacing with the external customer. Data is collected related to key external customer impact areas and compared with the way the organization is operating in those areas.

Conceptual clustering: This is the inherent structure of the data (concepts) that drives cluster formation. Since this technique is dependent on language, it is open to interpretation and consensus is required.

Concept tree (see conceptual clustering).

Confirmation bias: *The tendency of people to include or exclude data that does not fit a preconceived position.*

Consumer co-creation: *This means fostering individualized interactions and experience outcomes between a consumer and the producer of the organizational output. This can be done throughout the whole product life cycle. Customers may share their needs and comments, and even help spread the word or create communities in the commercialization phase. This approach provides a one-time limited interaction with consumers. Today, it is possible to enable constant interactions to really transfer knowledge, needs, desires, and trends from the consumer in a more structured way: co-creation.*

Contingency planning: *A process that primarily delivers a risk management strategy for a business to deal with the unexpected events effectively and the strategy for the business recovery to the normal position. The output of this process is called "contingency plan" or "business continuity and recovery plan."*

Contradiction analysis: The process of identifying and modeling contradictory requirements within a system, which, if unresolved, will limit the performance of the system in some manner.

Contradictions: TRIZ defines two kinds of contradictions: physical and technical.

Convergent thinking: Vetting the various ideas to identify the best workable solutions.

Copyrights: Legal protection of original works of artistic authorship.

Core or line extensions, renovation, sustaining close-in innovation: Extends and adds value to an existing line or platform of products via size, flavor, or format. It is incremental improvement to existing products.

Cost–benefit analysis (CBA): *A financial analysis where the cost of providing (producing) a benefit is compared with the expected value of the benefit to the customer, stakeholder, etc.*

Counseling and mentoring: These people love teaching, coaching, and mentoring. They like to guide employees, peers, and even their clients to better performance.

Crazy quilt principle: This is based on the expert entrepreneur's strategy to continuously seek out people who may become valuable contributors to his or her venture.

Create: To make something; to bring something into existence. The difference between creativity and innovation is that the output from innovation has to be a value-added output, while the output from creativity does not have to be value added.

Creative (preferred definition): Using the ability to make or think of new things involving the process by which new ideas, stories, products, etc., are created.

Creative problem solving (CPS): *A methodology developed in the 1950s by Osborn and Parnes. The method calls for solving problems in sequential stages with the systematic alternation of divergent and convergent thinking. It can be enhanced by the use of various creative tools and techniques during different stages of the process.*

Creative production: These people love beginning projects, making something original, and making something out of nothing. This can include processes or services as well as tangible objects. They are most engaged when inventing unconventional solutions. In an innovation process, these people may thrive on the ideation phase, creating multiple solutions to the identified problems.

Creative thinking: Creative thinking is all about finding fresh and innovative solutions to problems, and identifying opportunities to improve the way that we do things, along with finding and developing new and different ideas. It can be described as a way of looking at problems or situations from a fresh perspective that suggests unorthodox solutions, which may look unsettling at first.

Creativity: Creativity is the mental ability to conceptualize or imagine new, unusual, or unique ideas, to see the new connection between seemingly random or unrelated things.

Cross-industry innovation: This refers to innovations stemming from cross-industry affinities and approaches involving transfers from one industry to another.

Crowdfunding: The collective effort of individuals who network and pool their money, usually via the Internet, to support efforts initiated by other people or organizations.

Crowdsourcing: A term for a varied group of methods that share the attribute that they all depend on some contribution from the crowd. According to Howe, it is the act of a company or institution taking a function once performed by employees and outsourcing it to an undefined (and generally large) network of people in the form of an *open call.*

Culture: Culture is all about how people behave, treat each other, and treat customers.

Culture creator: Ensures the spirit of innovation is understood, celebrated, and aligned with the strategy of the organization.

Customer advocate: Keeps the voice of the customer alive in the hearts, minds, and actions of innovators and teams.

Customer profile: Empathy map is a technique for creating a profile of your customer beyond the simple demographics of age, gender, and income that has been in use for some time.

DVF model (desirable, viable, feasible): Another name for the balanced breakthrough model.

Design innovation: This focuses on the functional dimension of the job to be done, as well as the social and emotional dimensions, which are sometimes more important than functional aspects.

Design for X (DFX): Both a philosophy and methodology that can help organizations change the way that they manage product development and become more competitive. DFX is defined as a knowledge-based approach for designing products to have as many desirable characteristics as possible. The desirable characteristics include quality, reliability, serviceability, safety, user-friendliness, etc. This approach goes beyond the traditional quality aspects of function, features, and appearance of the item.

Design of experiments: This method is a statistically based method that can reduce the number of experiments needed to establish a mathematical relationship between a dependent variable and independent variables in a system.

Directed innovation: Directed innovation is a systematic approach that helps cross-functional teams apply problem-solving methods like brainstorming, TRIZ, creative problem solving, Six Thinking Hats™, Lateral Thinking™, assumption storming, inventing, Question Banking™, and provocation to a specifically defined problem in order to create novel and patentable solutions.

Direction setter: This creates and communicates vision and business strategy in a compelling manner, and ensures innovation priorities are clear.

Disruptive innovation: A process where a product or service takes root initially in simple applications at the bottom of a market and then relentlessly moves upmarket, eventually displacing established competitors.

Divergent thinking: Coming up with many ideas or solutions to a problem.

Diversity trumps ability theorem: This theorem states that a randomly selected collection of problem solvers outperforms a collection of the best individual problem solvers.

Drive to acquire: The drive to acquire tangible goods such as food, clothing, housing, and money, but also intangible goods such as experiences, or events that improve social status.

Drive to bond: The need for common kinship bonding to larger collectives such as organizations, associations, and nations.

Drive to comprehend: People want to be challenged by their jobs, to grow and learn.

Drive to defend: This includes defending your role and accomplishments. Fulfilling the drive to defend leads to feelings of security and confidence.

Edison method: The Edison method consists of five strategies that cover the full spectrum of innovation necessary for success.

Effectuation: Taking action toward unpredictable future states using currently controlled resources and with imperfect knowledge about current circumstances.

Ekvall: Ekvall's model of the creative climate identifies ten factors that need to be present:
- Idea time
- Challenge
- Freedom
- Idea support
- Conflicts
- Debates
- Playfulness, humor
- Trust, openness
- Dynamism
- Liveliness

Elevator speech: *An elevator speech is a clear, brief message or "commercial" about the innovative idea you are in the process of implementing. It communicates what it is, what you are looking for, and how it can benefit a company or organization. It is typically no more than two minutes, or the time it takes people to ride from the top to the bottom of a building in an elevator.*

Emergent collaboration: A social network activity where a shared perspective emerges from a group through spontaneous (unplanned) interactions.

Emotional rollercoaster: It is a notion, similar to journey mapping, that identifies areas of high anxiety in a process and, as such, exposes opportunities for new solutions.

Enterprise control: These people love to run projects or teams and control the assets. They enjoy owning a transaction or sale, and tend to ask for as much responsibility as possible in work situations.

Entrepreneur: Someone who exercises initiative by organizing a venture to take benefit of an opportunity and, as the decision maker, decides what, how, and how much of a good or service will be produced. An entrepreneur supplies risk capital as a risk taker, and monitors and controls the business activities. The entrepreneur is usually a sole proprietor, a partner, or the one who owns the majority of shares in an incorporated venture. From the business dictionary.com.

Era-based questions: Era-based questions require that you put yourself in the position of thinking about a question in a different time or place from the one you are currently in.

Experiments: In this context, experiments represent a mixture of surveys and observations in an artificial setting and can be summarized as test procedures.

Ethnography: *Ethnography can be used in many ways, but most significantly in the creation of a new product or service with a clear understanding of the many different ways that a person may accomplish a task based on their own world view. It means observing and recording what people do to solve a problem and not what they say the problems are. It is based on anthropology but used on current human activities. It is based on the belief that what people do can be more reliable than what they say.*

FAST: *An innovative technique to develop a graphical representation showing the logical relationships between the functions of a project, product, process, or service based on the questions "How" and "Why." In this case, it should not be confused with FAST that stands for the Fast Action Solution Team methodology created by H.J. Harrington. The two are very different in application and usage.*

Failure mode effects analysis: A matrix-based method used to investigate potential serious problems in a proposed system prior to final design. It creates a risk priority number that can be used to create a ranking of the biggest risks and then ranks the proposed solution.

Financial management: *Activities and manages financial programs and operations, including accounting liaison and pay services; budget preparation and execution; program, cost, and economic analysis; and nonappropriated fund oversight. It is held responsible and accountable for the ethical and intelligent use of investors' resources.*

Financial reporting: *Includes the main financial statements (income statement, balance sheet, statement of cash flows, statement of retained earnings, statement of stockholders' equity) plus other financial information such as annual reports, press releases, etc.*

Fishbone diagram, also known as Ishikawa diagram: A mnemonic diagram that looks like the skeleton of a fish and has words for the major spurs that prompt causes for the problem.

Five dimensions of a service innovation model:
- Organizational
- Product
- Market
- Process
- Input

Flowcharting: *A method of graphically describing an existing or proposed process by using simple symbols, lines, and words to pictorially display the sequence of activities. Flowcharts are used to understand, analyze, and communicate the activities that make up major processes throughout an organization. It can be used to graphically display the movement of product, communications, and knowledge related to anything that takes an input and value to it and produces an output.*

Focus group: *It is made up of a group of individuals that are knowledgeable or would make use of the subject being discussed. The facilitator is used to lead the discussions and record key information related to the discussions.*

Focus group: A focus group is a structured group interview of typically seven to ten individuals who are brought together to discuss their views related to a specific business issue. The group is brought together so that the organizer gain information and insight into a specific subject or the reaction to a proposed product.

Force field analysis: *A visual aid for pinpointing and analyzing elements that resist change (restraining forces) or push for change (driving forces). This technique helps drive improvement by developing plans to overcome the restrainers and make maximum use of the driving forces.*

Four dimensions of innovation:
- Technology: technical uncertainty of innovation projects
- Market: targeting of innovations on new or not previously satisfied customer needs
- Organization: the extent of organizational change
- Innovation environment: impact of innovations on the innovation environment

Four square model: The four-square model is a design process that consists of five steps:
- Problem framing: Identify what problem we intend to solve and outline a general approach for how we will solve it.
- Research: Gather qualitative and quantitative data related to the problem frame.
- Analysis: Unpack and interpret the data, building conceptual models that help explain what we found.
- Synthesis: Generate ideas and recommendations using the conceptual model as a guide.
- Decision making: Conduct evaluative research to determine which concepts or recommendations best fit the desirable, viable, and feasible criteria.

Four-square model for design innovation: Composed of two sets of polar extremes: understand–make and abstract–concrete.

Functional analysis: A standard method of systems engineering that has been adapted into TRIZ. The subject–action–object method is most frequently used now. It is a graphical and primarily qualitative methodology used to focus the problem solver on the functional relationships (good or bad) between system components.

Functional model: A structured representation of the functions (activities, actions, processes, operations) within the modeled system or subject area.

Functional innovation: Involves identifying the functional components of a problem or challenge and then addressing the processes underlying those functions that are in need of improvement. Through this process, overlaps, gaps, discontinuities, and other inefficiencies can be identified.

Futurist: Looks toward the future, scouts new opportunities, helps everyone see their potential. Enables people throughout the organization to discover the emerging trends that most impact their work.

Generic creativity tools: *A set of commonly used tools that are designed to assist individuals and groups to originate new and different thought patterns. They have many common characteristics like thinking positive, not criticizing ideas, thinking out of the box, right brain thinking, etc. Some of the typical tools are benchmarking, brainstorming, six thinking hats, storyboarding, and TRIZ.*

Goal: The end toward which effort is directed: the terminal point in a race. These are always specified in time and magnitude so they are easy to measure. Goals have key ingredients. First, they specifically state the target for the future state and second, they give the time interval in which the future state will be accomplished. These are key input to every strategic plan.

Goal-based questions: These questions pose the end goal without specifying the means or locking you into particular attributes.

Go-to-market investment: This is the cost of slotting fees for distribution, trade spending, advertising dollars (creative development, media spend), promotional programs, and digital/social media.

Gupta's Einsteinian theory of innovation (GETI): This theory states: Thus, every idea must have some energy associated with it that is an outcome of effort and the speed of the thought. Expressed as

- Innovation value = resources × (speed of thought)^2
- where the speed of thought can be described by the following relationship:
- Speed of thought ≡ function (knowledge, play, imagination)

HU diagrams: An effective way of providing a visual picture of the interface between harmful and useful characteristics of a system or process.

Hitchhiking: When a breakthrough occurs, it is a fertile area for innovators. They should hitchhike on the breakthrough to create new applications and improvements that can be inventions.

I-TRIZ: *An abbreviation for ideation TRIZ, which is a restructuring and enhancement of the classical TRIZ methodology based on modern research and practices. It is a guided set of step-by-step questions and instructions that aid teams in approaching, thinking, and dealing with systems targeted for innovation. It provides specific practical team guidance for the following applications:*

- *Solving a nontechnical or business issue*
- *Solving a technical or engineering issue*
- *Finding the root cause(s) of a system issue*

- *Anticipating and preventing possible systems*
- *Predicting and inventing specific innovative products or services customers will want in the future*
- *Patent (invention) evaluation, preparation, and enhancement to either work-around (invent around, design around) an existing (blocking) patent or provide a patent "fence" to prevent possible work-around*

Ideal final result (IFR): This states that in order to improve a system or process, the output of that system must improve (i.e., volume, quantity, quality, etc.), the cost of the system must be reduced, or both. It is an implementation-free description of the situation after the problem has been solved.

Idea priority index: Prioritizes ideas based on the potential cost–benefit analysis, associated risks, and likely time to commercialize the idea.

Idea selection by grouping or tiers: Groups can be helpful in evaluation of tiers like top ideas or worst ideas. Both grouping and tiers are only useful in a batch evaluation process, not a continuous process.

Idea selection by checklist or threshold: An individual idea's list of attributes must match the preset checklist or threshold in order to pass (e.g., be implemented in six months, profit of at least $500,000, and require no more than two employees).

Idea selection by personal preference: A manager, director, line employee, or even an expert is used to screen an idea on the basis of his or her own preferences.

Idea selection by point scoring: Uses a scoring sheet to rate a particular idea on its attributes (e.g., an idea that can be implemented in six months gets +5 points, and one that can make more than $\times$ dollars gets +10 points). The points are then added together, and the top ideas are ranked by highest total point scores.

Idea selection by priority index (IPI): The IPI prioritizes ideas based on the potential cost–benefit analysis, associated risks, and likely time to commercialize the idea, using the following relationships:

- Annualized potential impact of the idea = ($) $\times$ probability of acceptance
- IPI = annualized cost of idea development ($) $\times$ time to commercialize (years)

Idea selection by voting: Individual(s) can vote openly or in a closed ballot (i.e., blind or peer review). Voting can be weighted or an individual, such as expert, can give multiple votes to a given idea.

Image board, storyboarding, role playing: These are collections of physical manifestations (image collages or product libraries) of the desirable (or undesirable if you are using that as a motivator) to help generate ideas, or to facilitate conversations with users about what they want.

Imaginary brainstorming: It expands the brainstorming concept past the small group problem-solving tool to an electronic system that presents the problem/opportunity to anyone who is approved to participate in the electronic system. Creative ideas are collected, and a smaller group is used to analyze and identify innovative, imaginative concepts.

Indexing: Providing a tag for a fact, piece of information, or experience, so that you can retrieve it when you want it.

Influence through language and ideas: These people love expressing ideas for the enjoyment of storytelling, negotiating, or persuading. This can be in written or verbal form, or both.

Innovation: An advancement that transcends a limiting situation within the system under analysis. Another way to describe these limiting situations is to refer to them as contradictory requirements within a system.

Innovation: Converting ideas into tangible products, services, or processes. The challenge that every organization faces is how to convert good ideas into profit. That is what the innovation process is all about.

Innovation (preferred definition): The process of translating an idea or invention into an intangible product, service, or process that creates value for which the consumer (the entity that uses the output from the idea) is willing to pay for it more than the cost to produce it.

Innovation benchmarking: Comparing one organization, process, or product to another that is considered a standard.

Innovation blueprint: A visual map to the future that enables people within an enterprise or community to understand where they are headed and how they can build that future together. The blueprint is not a tool for individual innovators or teams to improve a specific product or service or to create new ones. Rather, the innovation blueprint is a tool for designing an enterprise that innovates extremely effectively on an ongoing basis.

Innovation culture: A culture that requires continuous learning, practices, and exceptions of risk and failure; holds individuals accountable for an action; and has aggressive timing.

Innovation management: The collection of ideas for new or improved products and services and their development, implementation, and exploitation in the market.

Innovation master plan framework: *The innovation master plan framework consists of five major elements: strategy, portfolio, processes, culture, and infrastructure.*

Innovation metrics: Measurements to validate that the organization innovate. They typically are

- Annual R&D budget as a percentage of annual sales
- Number of patents filed in the past year
- Total R&D headcount or budget as a percentage of sales
- Number of active projects
- Number of ideas submitted by employees
- Percentage of sales from products introduced in the past × year(s)

Innovation process: The innovation process is made up of five phases:

- Phase I: Creation phase
- Phase II: Value proposition phase
- Phase III: Resourcing phase
- Phase IV: Documentation phase
- Phase V: Production phase
- Phase VI: Sales/delivery phase
- Phase VII: Performance analysis phase

Innovative categories: Most service innovations can be categorized into one of the following groups:

- Incremental or radical, based on the degree of new knowledge
- Continuous or discontinuous, depending on its degree of price performance improvements over existing technologies. Sometimes called evolutionary innovation.
- Sustaining or disruptive, relative to the performance of the existing products
- Exploitative or evolutionary, innovation in terms of pursuing new knowledge and developing new services for emerging markets

Innovative problem solving: A subset of problem solving in that a solution must resolve a limitation in the system under analysis in order to be an innovative solution.

Innovator: An individual who creates a unique idea that is marketable and guides it through the innovative process so that its value to the customer is greater than the resources required to produce.

Insight: A linking or connection between ideas in the mind. The connections matter more than the pieces.

Inspiration: The word inspiration is from the Latin word *inspire*, meaning *to blow into*.

Inspire innovation tools: Tools that stimulate the unique creative powers. Some of them are

- Absence thinking
- Biomimicry
- Concept tree
- Creative thinking
- Ethnography
- HU diagrams
- Imaginary brainstorming
- I-TRIZ
- Mind mapping
- Open innovation
- Storyboarding
- TRIZ

Integrated innovation system: Covers the full end-to-end innovation process, and ensures the practices and tools are aligned and flow easily from one to the other.

Intellectual property rights: *The expression "intellectual property rights" refers to a number of legal rights that serve to protect various products of the intellect (i.e., "innovations"). These rights, while different from one another, can and do sometimes offer overlapping legal protection.*

Intersection of different sets of knowledge: Networking creates different relationships to be built across knowledge frontiers and opens up the participating organizations to new stimuli and experience.

Intrapreneur: An intrapreneur is an employee of a large corporation who is given freedom and financial support to create new products, services, systems, etc., and does not have to follow the corporation's usual routines or protocols.

Joint risk taking: Since innovation is a highly risky activity, it is very difficult for a single firm to undertake it by itself, and this impedes the development of new technologies. Joint collaboration minimizes the risk for each firm and encourages them to engage in new activities. This is the logic behind many precompetitive consortia collaborations for risky R&D.

Journey map or experience evaluations: A diagram that illustrates the steps your customer(s) go through in engaging with your company, whether it is a product, an online experience, a retail experience, a service, or any combination of these.

Kano analysis: A pictorial way to look at customer levels of dissatisfaction and satisfaction to define how they relate to the different product characteristics. The Kano method is based on the idea that features can be plotted using axes of fulfillment and delight. This defines areas of must haves, more is better, and delighters. It classifies customer preferences into five categories.

- *Attractive*
- *One-dimensional*
- *Must be*
- *Indifferent*
- *Reverse*

Kepner Trego: This method is very useful for processes that were performing well and then developed a problem. It is a good step-by-step method that is based on finding the cause of the problem by asking what changed since the process was working fine.

Knowledge management (KM): A strategy that turns an organization's intellectual assets, both recorded information and the talents of its members, into greater productivity, new value, and increased competitiveness. It is the leveraging of collective wisdom to increase responsiveness and innovation.

Leadership metrics: Leadership metrics address the behaviors that senior managers and leaders must exhibit to support a culture of innovation.

Lemonade principle: Based on the old adage that goes, "If life throws you lemons, make lemonade." In other words, make the best of the unexpected.

Link between climate and organizational innovation: Nine areas need to be evaluated to determine this linkage:

- Challenge, motivation
- Freedom
- Trust
- Idea time
- Play and humor
- Conflicts
- Idea support

- Debates
- Risk taking

Live-ins, shadowing, and immersion labs: They are designed to resemble the retail or home environment and gather extensive information about product purchase or use. These laboratories are used to both test the known, launch new product, and to observe user behavior.

Lotus blossom: *This technique is based on the use of analytical capacities and helps generate a great number of ideas that will possibly provide the best solution to the problem to be addressed by the management group. It uses a six-step process.*

Managing people and relationships: Unlike counseling and mentoring people, these people live to manage others on a day-to-day basis.

Marketing research: *Can be defined as the systematic and objective identification, collection, analysis, dissemination, and use of information that is undertaken to improve decision making related to products and services that are provided to external customers.*

Market research tools: The following are typical marketing research tools:

- Analysis of customer complaints
- Brainstorming
- Contextual inquiry, empathic design
- Cross-industry innovation
- Crowdsourcing
- In-depth interview
- Lead user technique
- Listening-in technique
- Netnography
- Outcome-driven innovation
- Quality function deployment
- Sequence-oriented problem identification, sequential incident technique
- Tracking, panel
- Analytic hierarchy process
- Category appraisal
- Concept test, virtual concept test
- Conjoint analysis
- Store and market test
- Free elicitation
- Information acceleration

- Information pump
- Kelly repertory grid
- Laddering
- Perceptual mapping
- Product test, product clinic
- Virtual stock market, securities trading of concepts
  i. Zaltman metaphor elicitation technique
  ii. Customer idealized design
  iii. Co-development
  iv. Expert Delphi discussion
  v. Focus group
  vi. Future workshop
  vii. Toolkit

Matrix diagram (decision matrix): *A systematic way of selecting from larger lists of alternatives. They can be used to select a problem from a list of potential problems, select primary root causes from a larger list, or to select a solution from a list of alternatives.*

Medici effect: The book by this name describes the intersection of significantly different ideas that can produce cross-pollination of fields and create more breakthroughs.

Mentor: Coaches and guides innovation champions and teams.

Methodology merger: Each methodology brings with it certain strengths and weaknesses that serve to fulfill specific steps and activities represented on the problem-solving pathway. When combined together and properly utilized, these methodologies create a very effective and useful outcome.

Mind mapping: *An innovation tool and method that starts with a main idea or goal in the middle, and then flows or diagrams ideas out from this one main subject. By using mind maps, you can quickly identify and understand the structure of a subject. You can see the way that pieces of information fit together in a format that your mind finds easy to recall and quick to review. They are also called spider diagrams.*

Mini problem: One that is solved without introducing new elements. We have to understand resources, since the emphasis is on solving the problem without introducing anything new to the system.

Moccasins/walking in the customer's shoes: The moccasins approach is more often called *walking in the customer's shoes.* This activity allows members of the organizations to directly participate in the process that the potential customer is subjected to by physically playing the role of the customer.

Myers–Briggs (MBTI): This is a survey-style measurement instrument used in determining an individual's social style preference.

NSD: An abbreviation for *new service development.*

Network-centric approach: The network-centric approach is taught in colleges and based on collaborative brainstorming. The concept is that more minds are better than one at a given time.

Networker: Works across organizational boundaries to engage stakeholders, promotes connections across boundaries, and secures widespread support.

Nominal group technique: A technique for prioritizing a list of problems, ideas, or issues that gives everyone in the group or team equal voice in the priority setting process.

Nonalgorithmic interactions: Actions with cognitive and physical materials of a project whose results you cannot predict for certain; those results you do not know.

Nonprobability sampling techniques: Use samples drawn according to specific and considered characteristics and are therefore based on the researcher's subjective judgment.

Nonprofit: An organization specifically formed to provide a service or product on a not-for-profit basis as determined by applicable law.

Observation: In this context, observation means the recording of behavioral patterns of people, objects, and events in order to obtain information.

Online collaboration: Convening an online brainstorming or idea generation session so members can participate remotely, instead of organizing a group in a room together,

Online management platforms: These are used to foster innovation and enable large groups of people to innovate together—across geographies and time zones. Users can post ideas and value propositions online and can collaborate with others to make these stronger. The community can rate and rank ideas or value propositions, post comments and recommendations, link to resources, build on each other's ideas, and support each other to improve each other's innovations.

Open innovation: The use and application of collective intelligence to produce a creative solution to a challenging problem, as well as to organize large amounts of data and information. The term refers to the

use of both inflows and outflows of knowledge to improve internal innovation and expand the markets for external exploitation of innovation. The central idea behind open innovation is that, in a world of widely distributed knowledge, companies cannot afford to rely entirely on their own research, but should instead buy or license processes or inventions (i.e., patents) from other companies.

Opportunity-driven model: Opportunity-driven model is more representative of street-smart individuals who take an idea at the right time and the right place, devise a solution, know how to market it, and capitalize on their breakthrough. They also appear to be lucky.

Organizational capability metrics: Organizational capability metrics focus on the infrastructure and process of innovation. Capability measures provide focus for initiatives geared toward building repeatable and sustainable approaches to invention and reinvention.

Organizational change management (OCM): *A comprehensive set of structured procedures for the decision-making, planning, executing, and evaluation activities. It is designed to minimize the resistance and cycle time to implementing a change.*

Organizational effectiveness measurements: The following is a typical way of measuring the organization's innovation effectiveness. Typically, it is measured in four key areas of management processes: product, sales, internal services, and sales and marketing. Each area is typically evaluated in a combination of the following:

- Committed leadership
- Clear strategy
- Market insights
- Creative people
- Innovative culture
- Competitive technologies
- Effective processes
- Supportive infrastructure
- Managed projects

Organization internal boundaries: Employee silos often isolate chains of command and communication, which can impede the progress of a valuable idea through product development.

Osborn method: Original brainstorming method developed by Alex F. Osborn by primarily requiring solicitation of unevaluated ideas (divergent thinking), followed by convergent organization and evaluation.

Outcome-driven innovation (ODI): *Built around the theory that people buy products and services to complete tasks or jobs they value. As people complete these jobs, they have certain measurable outcomes that they are attempting to achieve. It links a company's value creation activities to customer-defined metrics. Included in this method is the opportunity algorithm, which helps designers determine the needs that satisfied customers has. This help determine which features are most important to work on. Most important is this tool's intention of trying to find unmet needs that may lead to new and innovative products/services.*

PESTEL frameworks: The PESTEL framework focuses on the macroeconomic factors that influence a business. These factors are

- Political factors
- Economic factors
- Social factors
- Technological factors
- Environmental factors

Patents: A government-granted right that literally and strictly permits the patent owner to prevent others from practicing the claimed invention.

Performance engine project: A project that seeks to improve a current level of performance and not to create a new value proposition.

Permeability to innovation idea sources: Information and idea seeking differs greatly among companies.

Physical contradictions: Situations where one requirement has contradictory, opposite values to another.

Pilot in the plane principle: On the basis of the concept of *control*, using effectual logic, and is referred to as *nonpredictive control*. Expert entrepreneurs believe they can determine their individual futures best by applying effectual logic to the resources they currently control.

Pipeline model: Pipeline model, as driven by chance or innate genius, is a somewhat common perception of the innovation process. Inventors who work in research drive the pipeline model and development environment on a specific topic, explore new ideas, and develop new products and services.

Plan–do–check–act (PDCA): *A structured approach for the improvement of services, products, and/or processes. It is also sometimes referred to as plan–do–check–adjust. Another version of this PDCA cycle is OPDCA. The added "O" stands for observation or, as some versions say, "Grasp the current condition."*

Platform: A consumer need–based opportunity that inspires multiple innovation ideas with a sustainable competitive advantage to drive growth.

Portfolio management: The ongoing management of innovation to ensure delivery against stated goals and innovation strategy.

Post-Fordist: Companies after the Henry Ford efficient production era where managers wielded inordinate responsibility for profit and loss, and the new postmodern leaders of the global economy, who are responsible for developing talented teams.

Potential investor presentation: A short PowerPoint presentation designed to convince an individual or group to invest their money in an organization or a potential project. It can be a presentation to an individual or group not part of the organization, or the management of the organization that the presenter is presently employed by. It is usually part of a short meeting that usually lasts no more than 1 hour.

Practices: To look at all the inputs that we have available for selection and all the available operations or routines that we can perform on those inputs, then to select those inputs and operations that will give us our desired results.

Primary data: Data collected from the field or expected customer.

Primary sources: Gathered directly from the source; for instance, if new customer opinions were required to justify a new product, then customer interviews, focus groups, or surveys would suffice.

Principles of invention: A set of 40 principles from a variety of fields such as software, health care, electronics, mechanics, ergonomics, finance, marketing, etc., used to solve problems.

Proactive personal creativity: Proactive strategies to be especially effective in increasing the originality and effectiveness of personal creativity:

- Self-trust
- Open up
- Clean and organize
- Make mistakes
- Get angry
- Get enthusiastic
- Listen to hunches
- Subtract instead of adding
- Physical motion

- Question the questions
- Pump up the volume
- Read, read, read

Probability sampling techniques: Use samples randomly drawn from the whole population.

Probe-and-learn strategy: Where nonworking prototypes are developed in rapid succession, tested with potential customers, and feedback is sought on each prototype.

Problem detection and affinity diagrams: Focus groups, mall intercepts, or mail and phone surveys that ask customers what problems they have. They are all forms of problem detection. The responses are grouped according to commonality (affinity diagrams) to strengthen the validity of the response. Developing the correct queries and interpreting the responses are critical to the usefulness of the method.

Problem solving: Generating a workable solution.

Process: A series of interrelated activities or tasks that take an input and produces an output.

Process phases:
- Phase 1: Opportunity identification
- Phase 2: Idea Generation
- Phase 3: Concept evaluation
- Phase 4: Acquiring resources
- Phase 5: Development
- Phase 6: Producing the product
- Phase 7: Launch
- Phase 8: Sales and marketing
- Phase 9: Evaluation of results

Process innovation: Innovation of internal processes. New or improved delivery methods may occur in all aspects of the supply chain in an organization.

Process redesign: A methodology used to streamline a current process with the objective of reducing cost and cycle time by 30% to 60% while improving output quality from 20% to 200%.

Process reengineering: A methodology used to radically change the way a process is designed by developing an aggressive vision of how it should perform and using a group of enablers to prepare a new process design that is not hampered by the present processes paradigms. Use when a 60% to 80% reduction in cost or cycle time is required. Process reengineering is sometimes referred to as new process design or process innovation.

Product innovation: A multidisciplinary process usually involving many different functions within an organization and, in large organizations, often in coordination across continents.

Project management: *The application of knowledge, skills, tools, and techniques to project activities in order to meet or exceed stakeholders' needs and expectations from a project. (Source: PMBOK Guide.)*

Proof of concept (POC): *A demonstration, the purpose of which is to verify that certain concepts or theories have the potential for real-world application. A proof of concept is a test of an idea made by building a prototype of the application. It is an innovative, scaled-down version of the system you intend to develop. The proof of concept provides evidence that demonstrates that a business model, product, service, system, or idea is feasible and will function as intended.*

Pyramiding: A search technique in which the searcher simply asks an individual (the starting point) to identify one or more others who he or she thinks has higher levels of expertise.

Qualitative research (survey): Represents an unstructured, exploratory research methodology that makes use of psychological methods and relies on small samples, which are mostly not representative.

Qualitative research: Gathered data is transcribed, and single cases are analyzed and compared in order to find similarities and differences to gain deeper insights into the subject of interest. In *quantitative research*, the data preparation step contains the editing, coding, and transcribing of collected data.

Quality function deployment (QFD), also known as the house of quality: This creates a matrix that looks like a house that can mediate the specifications of a product or process. There are subsequent derivative houses that further mediate downstream implementation issues.

Quantitative analysis: These people love to use data and numbers to figure out business solutions. They may be in classic quantitative data jobs, but may also like building computer models to solve other types of business problems. These people can fall into two camps: (i) descriptive and (ii) prescriptive.

Quantitative research (survey): Can be seen as a structured research methodology based on large samples. The main objective in quantitative research is to quantify the data and generalize the results from the sample, using statistical analysis methods.

Quickscore creativity test: *A three-minute test that helps assess and develop business creativity skills.*

ROI metrics: ROI metrics address two measures: resource investments and financial returns. ROI metrics give innovation management fiscal discipline and help justify and recognize the value of strategic initiatives, programs, and the overall investment in innovation.

Radical innovation: A high level of activity in all four dimensions, while incremental innovations (low degree of novelty) are only weakly to moderately developed in the four dimensions.

Ranking or forced ranking: Ideas are ranked (#1, 2, 3, etc.)—this makes the group consider minor differences in ideas and their characteristics. For forced ranking, there can only be a single #1 idea, a single #2 idea, and so on.

Rating scales: An idea is rated on a number of preset scales (e.g., an idea can be rated on a one to ten on implementation time; any idea that reaches a nine or ten is automatically accepted).

Reverse engineering: *This is a process where organizations buy competitive products to better understand how the competitor is packaging, delivering, and selling their product. Once the product is delivered, it is tested, disassembled, and analyzed to determine its performance; how it is assembled; and to estimate its reliability. It is also used to provide the organization with information about the suppliers that the competitors are using.*

Robust design: *Robust design is more than a tool; it is complete methodology that can be used in the design of systems (products or processes) to ensure that they perform consistently in the hands of the customer. It comprises a process and tool kit that allows the designer to assess the impact of variation that the system is likely to experience in use, and if necessary redesign the system if it is found to be sensitive.*

Role model: Provides a living example of innovation through attention and language, as well as through personal choices and actions. Key stakeholders often test the leader's words, to see if these are real. For innovation to move forward, the leader must pass these inevitable tests—to show that, yes, he or she is absolutely committed to innovation as essential to success.

Root cause analysis (RCA): A graphical and textual technique used to understand complex systems and the dependent and independent fundamental contributors, or root causes, of the issue or problem under analysis. This is a technical process in that it provides specific direction as to how to execute the method.

Rote practice: Those activities where it looks like people are engaged in finding the right routines and inputs to obtain the desired result, but are just going through the motions.

S-curve: *A mathematical model also known as the logistic curve, which describes the growth of one variable in terms of another variable over time. S-curves are found in many fields of innovation, from biology and physics to business and technology.*

SCAMPER: *A tool that helps people to generate ideas for new products and services by encouraging them to think about how you could improve existing ones by using each of the six words that SCAMPER stands for, and applying it to the new product or service in order to generate additional new ideas. SCAMPER is a mnemonic that stands for*

- Substitute
- Combine
- Adapt
- Modify
- Put to another use
- Eliminate

SIPOC: An acronym for *supplier, input, process, output,* and *customer* model.

Scarcity of innovation opportunities: Markets have matured into commoditized exchanges.

Scenario analysis: *A process of analyzing possible future events by considering alternative possible outcomes (sometimes called "alternative worlds"). Thus, the scenario analysis, which is a main method of projections, does not try to show one exact picture of the future. Instead, scenario analysis is used as a decision-making tool in the strategic planning process in order to provide flexibility for long-term outcomes.*

Scientific method: The classical method that uses a hypothesis based on initial observations and validation through testing and revision if needed.

Secondary data: Data collected through in-house (desk research).

Secondary data sources: Involve evidence gathered from someone other than the primary source of the information. Most media outlets, magazines, books, articles, trade journals, market research reports, or publisher-based information are considered secondary sources of evidence.

Service innovation: Not substantially different than product innovation in that the goal is to satisfy customers' jobs-to-be-done, wow and retain customers, and ultimately optimize profit.

Seven key barriers to personal creativity: Seven key barriers to personal creativity:

- Perceived definitions of creativity
- Presumed uses for creativity
- Overdependence on knowledge
- Experiences and expertise
- Habits
- Personal and professional relationship networks
- Fear of failure

Simulation: *As used in innovation, simulation is the representation of the behavior or characteristics of one system through the use of another system, especially a computer program designed for the purpose. As such, it is both a strategy and a category of tools—and is often coupled with CAI (computer-aided innovation), which is an emerging simulation domain in the array of computer-aided technologies. CAI has been growing as a response to greater industry innovation demands for more reliability in new products.*

Six Sigma: A method designed for the reduction of variation in processes. The general steps used within the DMAIC (define, measure, analyze, improve, and control) and DMADV (define, measure, analyze, develop, and verify) methodologies are mostly administrative in nature. Combining Six Sigma with other tool sets pushes the process strongly toward the technical end of the scale.

Six Thinking Hats: *It is used to look at decisions from a number of important perspectives. This forces you to move outside your habitual thinking style, and helps you to get a more rounded view of a situation. The thinking is that if you look at a problem with the "Six Thinking Hats" technique, then you will solve it using any and all approaches. Your decisions and plans will mix ambition, skill in execution, public sensitivity, creativity, and good contingency planning.*

Social business: The practice of using social technologies to transform business.

Social innovation: Social innovation relates to creative ideas designed to address societal challenges—cultural, economic, and environmental issues—that are no longer simply a local or national problem but affect the well-being of the planet's inhabitants and ultimately, corporate profits and sustainability.

Social media: Refers to using social technologies as media in order to influence large audiences.

Social networks: *Networks of friends, colleagues, and other personal contacts: strong social networks can encourage healthy behaviors. They are often an online community of people with a common interest who use a website or other technologies to communicate with each other and share information, resources, etc. A business-oriented social network is a website or online service that facilitates this communication.*

Spontaneous order: A term that Hayek uses to describe what he calls the open society. It is created by unleashing human creativity generally in a way not planned by anyone, and, importantly, could not have been.

Stage gate process: First introduced by R.G. Cooper in 1986 in his book *Winning at New Products.*

Stakeholder: *A "stakeholder" of an organization or enterprise is someone who potentially or really influences that organization, and who is likely to be affected by that organization's activities and processes. Or, even more significantly,* **perceives** *that they will be affected (usually negatively).*

Statistical analysis: *A collection, examination, summarization, manipulation, and interpretation of quantitative data to discover its underlying causes, patterns, relationships, and trends.*

Storyboarding: *Physically structuring the output into a logical arrangement. The ideas, observations, or solutions may be grouped visually according to shared characteristics, dependencies on one another, or similar means. These groupings show relationships between ideas and provide a starting point for action plans and implementation sequences.*

Substantial platform, transformational, adjacencies innovation: Innovations that deliver a unique or new benefit or usage occasion, within an existing or adjacent category.

Synectics: *It combines a structured approach to creativity with the freewheeling problem-solving approach used in techniques like brainstorming. It is a useful technique when simpler creativity techniques like SCAMPER, brainstorming, and random input have failed to generate useful ideas. It uses many different triggers and stimuli to jolt people out of established mind-sets and into more creative ways of thinking.*

Systematic innovation stages: Systematic innovation can be viewed as occurring in stages:

- Concept stage
- Feasibility stage
- Development stage
- Execution stage, preparation for production
- Production stage
- Sustainability stage

Systematic innovation tools:

- Analogical thinking and mental simulations
- Theory of inventive problem solving (TRIZ)
- Scientific method
- Edison method
- Brainstorming
    i. Osborn method
    ii. Six Thinking Hats
    iii. Problem detection and affinity diagrams
    iv. Explore unusual results
    v. Ethnography
    vi. Function analysis and fast diagrams
    vii. Kano method
    viii. Abundance and redundancy
    ix. Hitchhiking
    x. Kepner Trego
    xi. Quality function deployment (QFD), also known as the house of quality
    xii. Design of experiments
    xiii. Failure mode effects analysis
    xiv. Fishbone diagrams, also known as Ishikawa diagrams
    xv. Five whys
    xvi. Medici effect
    xvii. Technology mapping and recombination
- Trial and error

System operator (also called *nine windows* or *multiscreen* method): A visual technique that is used frequently in the initial stages of TRIZ as part of problem definition.

System operator: The construction of a 3 × 3 matrix, with the rows labeled as the system, subsystem, super system; and the columns labeled past, present, and future.

Systems engineering: These methods are more technical than administrative processes, as they are fairly specific as to how to create and utilize the various systems engineering models. However, these methods guide the problem solver understanding that full system analysis is necessary in creating truly effective solutions. Therefore, these methods may be more administrative in nature than technical.

Systems engineering, system analysis: A technique to ensure that full system effects, impacts, benefits, and responses are understood when looking at changes or problems within a system.

Systems thinking: *An approach to problem solving, by viewing "problems" as parts of an overall system, rather than reacting to specific parts, outcomes, or events and potentially contributing to further development of unintended consequences.*

TEDOC methodology: TEDOC stands for target, explore, develop, optimize, and commercialize data to solve problems creatively. As such, TRIZ brings repeatability, predictability, and reliability to the problem-solving process with its structured and algorithmic approach.

TRIZ (pronounced "treesz"): *A Russian acronym for "Teoriya Resheniya Izobretatelskikh Zadatch," the theory of inventive problem solving, originated by Genrich Altshuller in 1946. It is a broad title representing methodologies, toolsets, knowledge bases, and model-based technologies for generating innovative ideas and solutions. It aims to create an algorithm to the innovation of new systems and the refinement of existing systems, and is based on the study of patents of technological evolution of systems, scientific theory, organizations, and works of art.*

Technical contradictions: The classical engineering and management trade-offs. The desired state cannot be reached because something else in the system prevents it. The TRIZ patent research classified 39 features for technical contradictions.

Technical process: Specifies not only what needs to be done, and in what order, but also provides specific details of *how* to execute the various tasks.

Technically focused brainstorming: The use of standard brainstorming methods bounded by certain acceptable solution concept conditions and guided by the attainment of an *ideal solution*.

Technically focused brainstorming: This methodology guides the generation of solution concepts by ensuring that those solution concepts support the resolution of contradictory requirements of the system under analysis, and renders that system to be of higher value than it was before the solution was applied.

Technology mapping and recombination: A matrix-based method that lists the various technologies that can perform a function and then examines combinations that have not been tried to see if there is enhanced performance or features.

Theory development and conceptual thinking: These people love thinking and talking about abstract ideas. They love the *why* of strategy more than the *how*. They may enjoy business models that explain the reasons behind the competitive position of a business.

Things are not innovation: The following maintenance or change management activities are *not* types of innovation:

- Cost savings
- Ingredient or product changes
- Regulatory change
- Label change

Thinking innovatively: Thinking innovatively is soliciting ideas from everyone, which is a challenge. There is a need for training people in asking questions, thinking of ideas, and articulating their ideas in words or graphics.

Thrashing: A term used to describe ineffective human workgroup activity, effort lost in unproductive work.

Time-of-day map: This tool focuses the participants not on a task or an experience, but rather what happens or does not happen in two- to four-hour chunks in a person's day and what opportunities may appear.

Top–down planning for innovation: Generally, a revenue goal-driven process that is usually set from the top by the senior leadership team. It is usually a dollar revenue goal or a percentage of revenue target from innovation.

Trademarks: Words or logos that are used by someone to identify their products or services, and distinguish them from the words or logos of others.

Trade secrets: Essentially refers to the legal protection often granted to confidential information having at least potential competitive value.

Tree diagram breakdown (drilldown): A technique for breaking complex opportunities and problems down into progressively smaller parts. Start by writing the opportunity statement or problem under investigation down on the left-hand side of a large sheet of paper. Next, write down the points that make up the next level of detail a little to the right of this. These may be factors contributing to the issue or opportunity, information relating to it, or questions raised by it. For each of these points, repeat the process. This process of breaking the issue under investigation into its component part is called drilling down.

Trends: Those dimensions on which lead users are far ahead of the mass market.

Trial and error: Attempts at successful solutions to a problem with little benefit from failed attempts.

Value analysis: The analysis of a system, process, or activity to determine which parts of it are classified as real-value-added (RVA), business-value-added (BVA), or no-value-added (NVA).

Value proposition: A document that defines the benefits that will result from the implementation of a change or the use of an output as viewed by one or more of the organization's stakeholders. A value proposition can apply to an entire organization, parts thereof, or customers, or products, or services, or internal processes.

Venture capitalist: Secures funding for innovation, evaluates and selects projects to receive resources, and guides implementation.

Vision: A documented or mental description or picture of a desired future state of an organization, product, process, team, a key business driver, activity, or individual.

Vision statements: A group of words that paints a clear picture of the desired business environment at a specific time in the future. Short-term vision statements usually are from three to five years. Long-term vision statements usually are from ten to 25 years. A visions statement should not exceed four sentences.

Zones of conflict: Refers to the temporal zone and the operating zone of the problem—loosely the time and space in which the problem occurs.

Index

Page numbers followed by f, n, and t indicate figures, notes, and tables, respectively.